OVERNIGHT LOAN
This book is to be returned on
or before the date stamped below

2002

0 4 MAY 2000

2 3 JAN 2003

2 4 JAN 2003

10.30am

2 3 APR 2004

- 5 MAY 2000

29 APR 2004

10.30am

2 3 MAY 2000

0 7 DEC 2004

2 3 MAY 2005

UNIVERSITY OF PLYMOUTH

PLYMOUTH LIBRARY

~~Short Loans cannot be renewed by phone~~
CHARGES WILL BE MADE FOR OVERDUE BOOKS

The Environmental Challenge
for Central European Economies
in Transition

Endorsements

Environmental issues are bound to play an important role in the upcoming discussions on enlargement of the European Union with new Member States from Central and Eastern Europe. This book is a very timely and competent contribution to these necessary discussions. As pointed out by the authors, environmental issues are not only a challenge for countries undergoing rapid social and political change, but also an opportunity. Written by experts from the countries concerned, the book helps the reader to appreciate better the nature and scope of the environmental problems to be faced in the region as well as the ways to overcome the problems. The book deserves to reach and be read by a wide audience, both in the present European Union countries and in the countries that are candidates for membership of the EU.

Ritt Bjerregaard
Commissioner for the Environment and Nuclear Safety,
European Commission

The Environmental Challenge for Central European Economies in Transition is a very timely and welcome book. The book is a concise, yet comprehensive and up-to-date compendium on progress achieved in environmental protection in the region of Central and Eastern Europe (CEE) since the revolutionary changes initiated in 1989. The book, written by top environmental experts who are well known for their important involvement in environmental protection activities in the region, focuses on environmental achievements during the six years of transition in CEE countries towards market-based, democratic societies. Reflection on the environmental situations of selected countries are coupled with fascinating overviews of attempts to transfer positive Western experiences in environmental management to the CEE region. The book also discusses the process of establishing the institutional capacity necessary to develop environmental protection policies and tools throughout the region.

Another strength of the book is that it examines the problems of integrating environmental protection into the transformation of societies and economies. The difficult question of how to assure economic recovery without compromising environmental restoration and conservation is not always positively answered, however, valuable experiences are shown. I strongly recommend this excellent publication for readers who want to know more about the state of the environment in CEE countries and better understand the complex processes leading to its gradual improvement.

Stanislaw Sitnicki
Executive Director, Regional Environmental Centre for
Central and Eastern Europe, Budapest, Hungary

The Environmental Challenge for Central European Economies in Transition

Edited by

Jürg Klarer
Regional Environmental Center, Budapest, Hungary (seconded by Swiss Government) and *Consultant to EAP Task Force, OECD*

and

Bedřich Moldan
Charles University, Prague, Czech Republic

JOHN WILEY & SONS
Chichester • New York • Weinheim • Brisbane • Singapore • Toronto

Other Wiley Editorial Offices

John Wiley & Sons, Inc., 605 Third Avenue,
New York, NY 10158-0012, USA

VCH Verlagsgesellschaft mbH, Papelalle 3, D-69469
Weinheim, Germany

Jacaranda Wiley Ltd, 33 Park Road, Milton,
Queensland 4064, Australia

John Wiley & Sons (Asia) Pte Ltd, 2 Clementi Loop #02-01,
Jin Xing Distripark, Singapore 129809

John Wiley & Sons (Canada) Ltd, 22 Worcester Road,
Rexdale, Ontario M9W 1L1, Canada

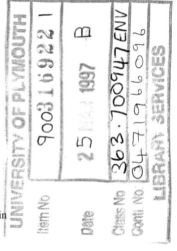

Library of Congress Cataloging-in-Publication Data

The environmental challenge for central European economies in transition / edited by Jürg Klarer and Bedřich Moldan.
    p.   cm.
Includes bibliographical references and index.
ISBN 0-471-96609-6 (acid-free paper)
1. Environmental policy — Europe, Eastern. 2. Environmental protection — Europe, Eastern. 3. Europe, Eastern — Environmental conditions. I. Klarer, Jürg. II. Moldan, Bedřich.
GE190.E852E585   1997
363.7′00947—dc20   96-43361
                                                                                      CIP

British Library Cataloguing in Publication Data

A catalogue record for this book is available from the British Library

ISBN 0-471-96609-6

Typeset in 10/12pt Times from authors' disks by Mayhew Typesetting, Rhayader, Powys
Printed and bound from postscript disk in Great Britain by Biddles Ltd, Guildford and King's Lynn

Contents

Contributors   ix

Preface   xi

CHAPTER 1   Regional Overview
          Jürg Klarer and Patrick Francis
1.1   DEVELOPMENTS AND TRENDS IN ENVIRONMENTAL
      PROTECTION IN CEE COUNTRIES IN TRANSITION   1
          The context   1
          Observations on selected achievements, problems and trends in
          environmental protection in CEE countries   19
1.2   THE EUROPEAN DIMENSION: TOWARDS
      CONVERGENCE OF ENVIRONMENTAL POLICIES   27
          The environment for Europe process   28
          European Union accession: implications for environmental policy 36
1.3   FINANCING ENVIRONMENTAL PROTECTION IN CEE
      COUNTRIES   44
          Impediments to financing environmental protection in CEE
          countries   46
          Domestic sources of environmental financing in CEE countries   50
          Foreign and international financing of environmental protection
          in CEE countries   55
          NOTES   63
          REFERENCES   63

CHAPTER 2   Bulgaria
          Kristalina Georgieva and Judith Moore
2.1   BACKGROUND   67
          Impact of the political and economic transition of the early
          1990s   68
          Past and current state of the environment   70
          Transboundary pollution issues   79
2.2   LEGAL AND ADMINISTRATIVE FRAMEWORK   84
          Legislative developments since 1990   86
          Implementing organizations   89

2.3 FINANCING ENVIRONMENTAL IMPROVEMENTS 92
Trends in environmental expenditures 92
Environmental funds 93
Trends in the allocation of resources 96
Foreign assistance 96
2.4 PUBLIC ATTITUDES TOWARDS THE ENVIRONMENT 99
An outlook on what people care about 99
Non-governmental organizations 100
2.5 CONCLUSIONS 102
REFERENCES AND FURTHER READING 105

CHAPTER 3 Czech Republic

Bedřich Moldan
3.1 BEFORE AND AFTER NOVEMBER 1989 107
3.2 ENVIRONMENTAL POLLUTION AND ITS
CAUSES 111
3.3 ENVIRONMENT AND SOCIETY 118
REFERENCES 129

CHAPTER 4 Hungary
Zsuzsa Lehoczki and Zsuzsanna Balogh
4.1 INTRODUCTION 131
4.2 PRESENT TRENDS IN THE STATE OF THE
ENVIRONMENT 132
Air pollution 133
Trends in energy use 137
Water pollution 139
Waste management 140
Nature conservation 144
4.3 INSTITUTIONS AND POLICIES: CHANGES IN THE
PUBLIC AND PRIVATE SECTORS 145
Changes in governmental tasks and responsibilities 145
Environmental policy 149
Changes in environmental legislation 151
Economic instruments 153
Privatization, bankruptcy procedures and the environment 158
Financing environmental protection 159
Private sector participation in environmental protection 161
Public perception and participation 163
4.4 CONCLUSIONS: CHALLENGES FOR ENVIRONMENTAL
PROTECTION 165
REFERENCES 167

CHAPTER 5   Latvia
   Valts Vilnitis
5.1   BACKGROUND   169
      Changes in society   169
      State of the environment   170
5.2   INSTITUTIONS AND POLICIES – KEY PLAYERS   180
      Existing environmental protection institutions and division of
      responsibility   180
      Influence of non-environmental institutions and sectors in
      decision-making   184
      Environmental policy   184
      Environmental law   185
      Relations with other sectors   187
      Education and information   187
      International relations and co-operation   188
5.3   PUBLIC PERCEPTION AND PARTICIPATION   189
      Public opinion about environmental issues   189
      State of public participation   190
5.4   MAJOR PROBLEMS AND OBSTACLES   191
      REFERENCES   192

CHAPTER 6   Poland
   Maciej Nowicki
6.1   PRESENT TRENDS IN ENVIRONMENTAL
      PROTECTION   193
      Air protection   194
      Water protection   197
      Waste management   200
      Nature conservation   203
6.2   KEY INSTITUTIONS   206
      The national level   206
      The regional level (voivodships)   209
      The local level (gminas)   209
      Agencies connected with environmental protection   210
6.3   TOOLS FOR ECOLOGICAL POLICY   211
      National environmental policy   211
      Reforms of the ecological law   212
      Financial instruments   213
      International co-operation   220
6.4   PARTICIPATION OF SOCIETY IN ENVIRONMENTAL
      PROTECTION   222
6.5   CONCLUSIONS   225
      REFERENCES AND FURTHER READING   226

CHAPTER 7 Slovak Republic
Mikuláš Huba
7.1 BACKGROUND 229
Brief characterization of Slovakia's environment 229
Some environmental statistics and trends 231
Impact of political and economic transformation 234
7.2 INSTITUTIONS AND POLICIES 239
Brief outline of the institutional set-up for environmental
protection 239
Key players 241
Environmental policy issues 247
7.3 PUBLIC PERCEPTION AND PARTICIPATION 255
Public opinion on environmental issues 255
Public participation in environmental decision-making 258
7.4 MAJOR PROBLEMS AND OBSTACLES 262
Major problems confronting environmental protection 262
The way ahead 265
REFERENCES 270

CHAPTER 8 Conclusions
Patrick Francis, Jürg Klarer and Bedřich Moldan
8.1 REFLECTIONS ON SIX CEE COUNTRIES 272
8.2 BROADER OBSERVATIONS ON THE ENVIRONMENTAL
CHALLENGES FACING CEE COUNTRIES 275

Index 281

Contributors

ZSUZSANNA DÓCSNÉ BALOGH is a consultant to the Ministry of Environment and Regional Policy of Hungary as an environmental economist. Previously she was with the Department of Economics and Budget Planning of the Ministry of Environmental Protection and Regional Policy.

PATRICK FRANCIS is an environmental finance specialist for the Environmental Action Programme Task Force at OECD. Previously he was Senior Adviser for the USAID sponsored 'Environmental Action Program Support Project', supporting environmental investments in Central and Eastern Europe. He has also been a consultant for the Harvard Institute for International Development in Poland, the Regional Environmental Centre for Central and Eastern Europe and was an adviser to Polish environmental organizations as a US Peace Corps Volunteer.

KRISTALINA GEORGIEVA is a Senior Environmental Economist in the Environment Division for Europe, Central Asia, Middle East and North Africa Regions of the World Bank. She is responsible for World Bank assistance to a broad range of environmental activities in Central and Eastern Europe and the former Soviet Union, including regional programmes, national environmental action plans and lending operations. She obtained her PhD in economics with a dissertation in environmental policy. Prior to joining the World Bank, Ms Georgieva was an Associate Professor at the Department of Economics of the University of National and World Economy in Sofia, Bulgaria.

MIKULÁŠ HUBA is a researcher on environmental issues at the Institute of Geography at the Slovak Academy of Sciences and is a leading figure of Slovakia's independent environmental movement. Previously Mr Huba was a Member of the Slovak Parliament and Chairman of the Parliamentary Committee for Environment (1990–92). He has also been Chairman of the Slovak Union of Nature and Landscape Protectors (SZOPK) and Friends of Earth Slovakia. Since 1993, Mikuláš Huba has been Chairman of the Society for Sustainable Living in Slovakia.

JÜRG KLARER is consultant to the Environmental Action Programme Task Force at OECD and Project Manager at the Regional Environmental

Center for Central and Eastern Europe (seconded by the Swiss Government). Specializing in environmental policy and environmental economics, Mr Klarer has also worked as a consultant for the Federal Department of Foreign Affairs and the Federal Office of Environment, Forests and Landscapes of the Swiss Government on Technical Environmental Assistance to Central and Eastern Europe, and the United Nations Economic Commission for Europe.

ZSUZSA LEHOCZKI is an environmental economist specializing in the use of economic instruments and new financing mechanisms for environmental protection. Currently Ms Lehoczki is Senior Policy Analyst for the Harvard Institute for International Development at the Hungarian Ministry of Finance in Budapest. She is also a Senior Lecturer at the Microeconomics Department of the Budapest University of Economic Sciences.

BEDŘICH MOLDAN is Director of the Charles University Environmental Centre in Prague and Associate Professor of Geochemistry. He is also Chairman of the Czech Union of Nature Conservation and Chairman of the Board of Directors of the Regional Environmental Centre for Central and Eastern Europe in Budapest. Previously Dr Moldan was a Member of the Czech Parliament (1990–92), Minister for the Environment of the Czech Republic 1989–91, the first ever Minister of Environment in Czechoslovakia, and in 1993–94 was Vice-Chairman of the United Nations Commission on Sustainable Development.

JUDITH MOORE is with the Department of Urban Studies and Planning, Massachusetts Institute of Technology, Cambridge, Massachusetts. Ms Moore has been a consultant and researcher on international environmental issues for several organizations including the World Bank, the World Wildlife Fund and the World Resources Institute.

MACIEJ NOWICKI is President of the 'EcoFund' Foundation, Poland's debt-for-environment swap mechanism, since its establishment in 1992 and was one of the initiators of this unique institution. In 1990–91, Professor Nowicki was Poland's Minister of Environmental Protection, Natural Resources and Forestry. He has also been Vice-Chairman of the United Nations Commission on Sustainable Development (1994–95).

VALTS VILNITIS is Director of the Environmental Protection Department of the Ministry of Environmental Protection and Regional Development of the Republic of Latvia. Previously Mr Vilnitis was Environmental Adviser to the Government of Latvia (1990–92), Executive Director of the Latvian Fund for Nature (1992), and Deputy Chairman of the Environmental Protection Committee of the Republic of Latvia (1992–93).

Preface

The countries of Central and Eastern Europe inherited some of the worst environmental problems in the world from more than 40 years of Communist rule and central planning. Their economies were several times more pollution intensive (on energy use and just about every dimension of pollution, air, water, waste) than the far richer economies of the West. Many 'hot spot' areas existed with extreme pollution loads and severe environmental degradation and risks for human health.

Since 1989, the revolutionary changes were initiated towards transition to democracy and market economy. There were high hopes that the countries of Central and Eastern Europe could make use of the 'window of opportunity' created by transition and introduce modern systems for environmental protection, but also learn from the mistakes and inefficiencies of environmental policy development in the West.

Today, more than six years of transition are completed and many of the most important steps initiated. High on the agenda of Central and Eastern European (CEE) countries is quick and full membership in Western structures, including membership in the European Union and OECD. In this book we analyse the results and impact the transition period brought so far for environmental protection in the context of the very dynamically changing economic, political and social circumstances.

Chapter 1 gives a regional overview in which similarities and differences in environmental protection development among the CEE countries are discussed, the pan-European dimension is outlined and environmental financing issues are analysed. Following chapters deal with specific countries.

Rather than covering all CEE countries with individual chapters, we focused on collecting high quality texts from the respective countries, written by respected environmental experts who have made strong personal and professional commitments to shaping the development of environmental protection in their country. The focus is on the four 'Visegrad' countries (Czech Republic, Hungary, Poland and Slovak Republic) which are most advanced in transition and for which the impact of transition can be analysed most comprehensively. The reader also can find a country chapter on Latvia, one of the three Baltic countries which gained full independence from the former Soviet Union. In addition, the book contains a chapter on Bulgaria, one of the Balkan countries of South-Eastern Europe. The country chapters follow a similar structure: changes in the state of the environment and policies, institutions and key actors for environmental protection. In the concluding chapter we summarize our main findings.

We would like to thank Sheila Crofut, US Peace Corps Volunteer in Prague, for her commitment and skilful work with language editing the country chapters and for her help with co-ordination.

Jürg Klarer and Bedřich Moldan
Prague, March 1996

CHAPTER 1

Regional Overview

Jürg Klarer and Patrick Francis

1.1 DEVELOPMENTS AND TRENDS IN ENVIRONMENTAL PROTECTION IN CEE COUNTRIES IN TRANSITION

The context

The revolutionary changes since 1989 in Central and Eastern European (CEE)[1] countries are of global importance. The reforms necessary following the dismantling of Communism are enormous. Entire economic, political and social systems are being re-built. 'Transformation', 'transition', 'reconstruction' and 'reform' are key words in the region. Economic recovery and growth based on free market economies have become the pre-eminent priorities of most CEE countries. The fundamental re-orientation of property rights in all areas including the privatization of enterprises and land are required in order to meet these goals. At the same time, people are demanding more individual rights, including the right to have a say in how their countries develop. New democracies are being built through 'top-down' legislation and policy and the 'bottom-up' demands and participation of individual citizens and independent organizations. Governments and enterprises, citizens and organizations are all faced with the task of creating civil societies, societies based on the rule of law, societies with functioning judicial systems which effectively protect basic human rights. Most CEE countries have also expressed the keen desire to become integrated into the mainstream of European development, to rejoin the Europe of which they used to be part. It is within this broader context that we examine the environmental challenge facing CEE countries as they endeavour to find their ways through the profound transformations taking place in the region.

CEE countries do not have to entirely 'reinvent the wheel'. If they so desire, they may draw upon the experience of Western and other countries as they design, implement and revise new political, economic and social systems.

The Environmental Challenge for Central European Economies in Transition.
Edited by J. Klarer and B. Moldan. © 1997 John Wiley & Sons Ltd.

Nevertheless, numerous, often profound challenges exist. To start with, today's CEE societies are lacking familiarity with the basic principles of democracy and market economies. Consequently, there is a huge ongoing learning (or better: learning-by-doing) process taking place in the region. Given the economic and social hardships the citizens are experiencing in the transition period, this learning process is of existential importance for many people in CEE countries. Moreover, tested and 'successful' Western systems cannot simply be copied and applied in CEE countries. Potential Western solutions must be adapted to the particular CEE societies, mentalities, and economies, and to the special circumstances during the transition period.

The first several years of transition show clearly that significant differences do exist between the CEE countries with respect to stage and nature of development and reform. Table 1.1 reflects some of these differences with factors affecting them including: depth and comprehensiveness of reform measures; speed of reform pursued; proximity to Western markets and trade; level of foreign investment in a country; and, underlying strength of a country's will and capacity to join the mainstream of Western economic development.

In most CEE countries, since 1994, the main economic indicators show positive and growing trends (see Table 1.2), and the process of institutionalizing and fine-tuning newly established systems has begun. At the same time, debates about further reforms as well as amendments to already realized measures are intensifying as the key actors and lobbies organize themselves better and 'play their role' more professionally. In some countries, however, institutional and policy reforms have progressed at a notably slower pace with further progress often characterized and hindered by struggles for power between fragmented new and old political interest groups.

The Legacy of the Past

CEE countries have interrelated but distinct histories. Their contemporary situations differ as well, as do their perspectives and strategies for the transformation process. However, among these countries a certain standardization in political and economic structures was imposed by more than 40 years of Soviet dominated Communist Party governments. Today, most of the major environmental problems are rooted in similar causes and structures inherited from the Communist systems.

The most important feature was central planning. The primary concern of Communist centralized planning was not the welfare of society, but rather the objectives of Soviet political ideology. Everything was directly or indirectly bound to that ultimate end, and the system was extremely wasteful and ultimately ineffective. A very obvious and direct consequence of central planning was the economic and industrial structure of CEE countries, which

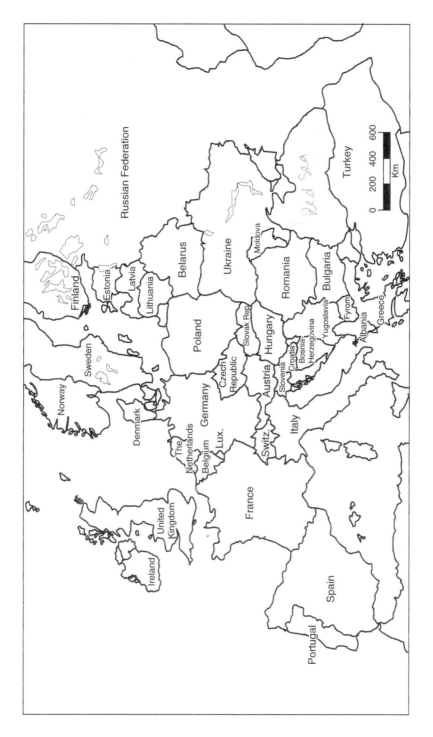

Figure 1.1   Map of CEE countries

Table 1.1 Basic country information

| | Total area 1994[1] (1000 km²) | Population mid-1994[1] (million) | Share of GDP in Sector (%) | | | | | | |
			mid-1995[2] Private sector	1990[3] Agriculture	1990[3] Industry	1990[3] Services	1994[1] (or latest available) Agriculture	1994[1] (or latest available) Industry	1994[1] (or latest available) Services
Czech Rep.	79	10.3	70	8	56	36	6	39	55
Hungary	93	10.3	60	12	32	56	7	33	60
Poland	313	38.5	60	14	36	50	6	40	54
Slovak Rep.	49	5.3	60	8	56	36	7	36	57
Bulgaria	111	8.4	45	18	52	31	13	35	53
Romania	238	22.7	40	18	48	34	21	33	46
Slovenia	20	2.0	45	na	na	na	5	38	57
Estonia	45	1.5	65	na	na	na	10	36	55
Latvia	64	2.5	60	na	na	na	9	34	57
Lithuania	65	3.7	55	na	na	na	21	41	38
Austria	84	8.0	…	3	37	60	2	34	64
France	552	57.9	…	4	29	67	2	28	70
UK	245	58.4	…	na	na	na	2	32	66
Portugal	92	9.9	…	na	na	na	na	na	na

Sources: [1]World Bank, 1996. [2]EBRD, 1995. (Data on private sector share are rough EBRD estimates.) [3]World Bank, 1992. (Data for Czech and Slovak Republic refer to Czechoslovakia.)

Table 1.2 Selected economic indicators

Country	GDP[1] (% change in period)		Industrial prod.[1] (% change in period)		Consumer prices[1] (% change in period)		Foreign direct investment (cumulative 1989–95)	
	1989–93	1994–95	1989–93	1994–95	1989–93	1994–95	mln US$[2]	per capita[3]
Czech Republic	−21	8	−35	12	230	19	3 996	388
Hungary	−18	4	−31	15	260	53	10 634	1 032
Poland	−13	12	−29	25	2 260	61	6 459	168
Slovak Republic	−24	12	−41	15	240	22	483	91
Bulgaria	−27	4	−52	9	1 720	161	397	47
Romania	−25	11	−53	13	3 190	195	1 101	49
Slovenia	−16	11	−34	9	5 630	30	438	219
Estonia	−35	9	na	na	8 530	91	646	431
Latvia	−48	−1	na	na	5 450	67	323	129
Lithuania	−61	4	na	na	20 090	133	73	20

Sources: [1]Author's calculations based on EBRD, 1994/5. [2]World Bank, 1996. [3]Author's calculations based on World Bank, 1996.

Notes: 1989 and 1990 data on GDP and Consumer Prices for Czech and Slovak Republics refer to CSFR. *Industrial production*: Hungary: gross industrial output; Slovakia: 1989–91 data cover state enterprises only, from 1992 private sector is included. †*Consumer prices*: Slovenia: data refer to retail prices.

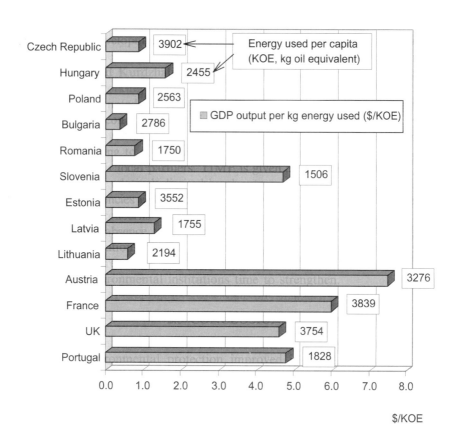

Figure 1.2   Energy intensities and energy use per capita in selected countries in 1994.
Source: World Bank, 1996

was dominated by oversized heavy industry. Steel, chemicals, mining, heavy machinery and energy were the dominant economic sectors in the Soviet sphere of influence, and the military–industrial complex enjoyed special priority. All kinds of resources were devoured by these industries including the better paid and the better educated workforce. One typical result of the ill-planned industrial structures is high and very inefficient consumption of energy, generated mainly from low quality fuels (see Figure 1.2).

Communist economies also lead to the excessive exploitation of natural resources of all sorts. The whole Soviet economy was based, first of all, on the enormous mineral richness of the former USSR which covered one-sixth of the world's land mass. The resources of CEE countries were exploited with equal brutality and recklessness. Even more so, because in CEE they were more easily accessible as compared to, for instance, the remote or even frozen

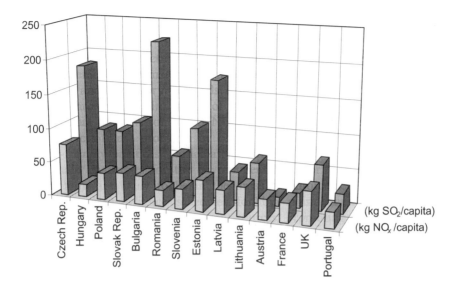

Figure 1.3 SO_2 and NO_x per capita emissions in selected countries in 1990. *Source*:
Eurostat, EC, EEA (1995). Based on data from CORINAIR

Siberia. As can often be seen in the legacy left behind, in the course of such exploitation virtually no environmental measures were applied.

With practically no exception, production processes were wasteful. The only important criteria were planned parameters, meaning production objectives (tonnes, numbers of products, gigajoules, etc.). There were no incentives to introduce efficient or environment-friendly technologies. In some countries, harmful agricultural and forestry practices based on central planning were also implemented. Production quotas set by bureaucrats in remote capitals were often fulfilled by excessive use of heavy machinery, chemicals and labour. The result of such processes was a degraded environment, impoverished people, suffering animals and food of poor quality, contaminated by a variety of harmful substances.

In Figures 1.3–1.5 and Table 1.3, some general indicators on the state of the environment in CEE countries are provided. These illustrations also contain comparative data for selected Western European countries. (Further analysis and more detailed information on the state of the environment can be found in the country chapters.)

Another legacy typically found in the CEE countries is Soviet-style urbanization based on concentrated, large urban settlements. In many cases residential settlements were located directly adjacent to large industrial (heavily polluting) complexes, while in others people were settled in drab apartment blocks, sometimes far from their workplaces, cultural centres or recreational facilities. These block settlements are still one of the most visible legacies of

kg/capita

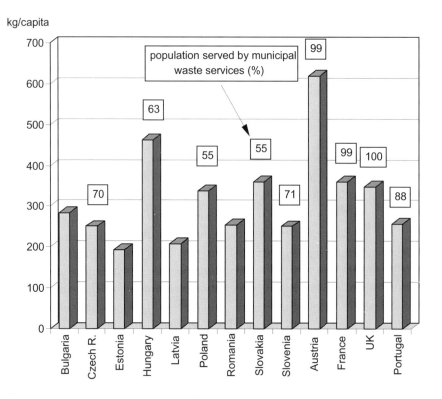

Figure 1.4   Municipal waste generation in selected countries (kg per capita, 1990).
Source: Eurostat, EC, EEA (1995)

the past regime. They were built not only in large cities, small towns and villages are also spoiled by them. Apart from being composed of almost identical ugly, grey concrete buildings throughout the region, these settlements have other features in common: almost no greenways; polluting central-heating plants; inadequate maintenance; insufficient wastewater and solid waste management facilities.

The neglect of environmental problems was pervasive throughout the system. It was exacerbated by general hypocrisies of the regime which pretended to be the most advanced, the most human, the most socially just, etc., where problems typical of 'rotten capitalism', e.g. environmental pollution, officially did not exist. The reluctance to admit the gravity of environmental problems went hand in hand with secretiveness. Anything not in line with the shining image of the regime was kept secret and, therefore, officially did not exist.

Blurred property rights was another feature of communist economic rules. Almost everything, especially all means of production, was in the 'peoples' ownership'. In practice, however, this meant ownership by nameless, faceless,

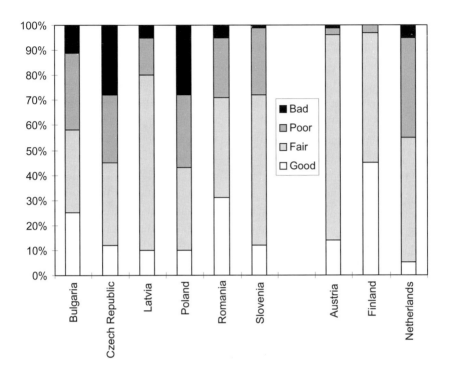

Figure 1.5 Quality of river reaches (1990). *Note*: Applied criteria for water quality estimation were: organic pollution and nutrient content; oxygen conditions; state of flora and fauna. *Source*: Europe's Environment – The Dobris Assessment, 1995

party–state bureaucrats. As a result, nobody was responsible for the state of any property, and cities, factories, houses and roads slowly but progressively deteriorated. This situation was aggravated by unrealistic pricing. No prices were set by market forces (indeed, no market existed), all were decided by bureaucrats. Signals of scarcity were scarcely available. Supplies of energy and raw materials were also heavily subsidized which led to even more squandering.

Since the 1970s, environmental legislation has already been developed and adopted in the CEE countries. However, this legislation was (and remains) incomplete, often not practically implementable, and simply not enforced. The laws were ostensibly genuine and standards strict, but they were neither enforced nor enforceable. In fact, the laws were never truly meant to be observed. They merely helped support the image of the system. There are several reasons why the laws were, in principle, not enforceable. First, the standards were unrealistically stringent without adequate technological or economic prerequisites. Second, the necessary institutional structure was

Table 1.3   Intensity of air pollution and per capita emissions in selected countries in 1993 (or latest available year)

	SO$_x$ emissions		NO$_x$ emissions		CO$_2$ emissions	
	(kg/capita)	(kg/1000 US\$ GDP)*	(kg/capita)	(kg/1000 US\$ GDP)*	(t/capita)†	(t/1000 US\$ GDP)*
Hungary	80.1	11.3	17.7	2.5	6.1	0.86
Poland	70.9	13.0	29.1	5.4	8.8	1.68
Bulgaria	168.1	33.7	28.1	5.6	6.1	1.19
Austria	8.9	0.5	22.8	1.3	7.1	0.41
France	21.3	1.2	26.5	1.4	6.4	0.36
UK	54.8	3.5	40.3	2.6	9.6	0.61
Portugal	29.0	2.8	21.9	2.2	4.6	0.45

Source: OECD, 1996.

Notes: *GDP data for Austria, France, UK and Portugal at 1991 prices and purchasing power parities (PPPs); GDP data for Bulgaria, Hungary and Poland are OECD Secretariat estimates based on 1993 PPPs. †CO$_2$ from energy use only; international marine bunkers are excluded.

lacking (no administrative bodies with real legal competence, no involvement of courts of justice, no adequate monitoring, etc.). Third, there was no adequate mechanism, and indeed no reason for it, for one body of the party–state bureaucracy to punish another body of the same bureaucracy if the law was not observed.

In all developed and democratic countries, the most important supporter of the environmental cause is the general public, the concerned citizens. Public participation is a decisive factor for efficient environmental protection. In a totalitarian Communist regime, no such thing was thinkable. In addition, there was an almost absolute lack of accurate information on the state of the environment. As environmental problems officially did not exist there was no education about such, and thus awareness and understanding of environmental issues was very low. Technological knowledge and skills were generally at a rather high level (comparable with the West) but expertise in other subjects like economics, managerial practices, the humanities, and environmental education was underdeveloped.

Especially in the initial phase after the dismantling of the Communist systems, however, the picture of environmental catastrophe in the CEE countries became widespread. Long-existing concern over the health effects caused by environmental pollution, especially in industrial areas, was freely and loudly expressed. As a result, environmental 'hot spots', often located in highly populated industrial areas, gained much attention and financial assistance. At the same time, certain features of the previous political and economic system, such as strict centralized authority and forced movements of some local populations, have left large areas of the CEE region untouched by industrial activity and sparsely inhabited. Pristine nature containing rich biodiversity is still estimated to cover about 30% of the area in the CEE countries (REC, 1994).

Experience from the transition process shows that attitudes, habits and mentalities formed under the Communist rule are often much harder to change than the economic and social settings in which they exist. In a number of CEE countries, traditions of the previous system live strong, and while institutions can be established rather quickly, political mentalities change slowly. Such realities continue to influence policy development in all areas, including environmental policy.

The Status of Environmental Issues in Political, Economic, and Social Transition

Political reform The environment was a high civil priority in the CEE countries on the eve of the political breakthrough, and was used by the public as a political weapon for showing the mistakes of the previous political and economic systems. Indeed, pressure from environmentalists contributed substantially to dismantling the previous political systems in most CEE countries. Environmental movements were sometimes the only civil movements which were more or less tolerated by the Communist supremacy. Therefore, such movements also attracted many people who used the environmental banner as an umbrella for protest actually motivated by other interests. Many of these people left the environmental movement after 1989 to pursue non-environmental careers (REC, 1994).

Nevertheless, in the first parliaments and governments of the new democracies, many prominent figures from such environmental movements were present and active, among them a high number of experts with high environmental awareness and knowledge of environmental issues. Today, however, after several re-elections in all CEE countries, the situation is different and fewer environmental enthusiasts are holding powerful, political positions. Environmental protection in the transition period turned out to be a complex and demanding task, and environmental issues quickly lost their appeal as a vehicle for a quick political career.

In the CEE region (as in the West), experience has shown that political debate on environmental issues is often not as constructive and solution-orientated as the topic requires. Another weakness of this debate is that the impacts of proposed actions on social welfare and economic development are often not afforded adequate consideration. Such discussions are commonly marked by the incorrect judgement and prejudice that environmental protection constitutes merely a drain on financial resources and an obstacle to economic development (REC, 1994).

Democratization of the political arena has altered the way power is exercised and the CEE countries are in the process of reforming their government structures. One important element of reform is decentralization. Independent self-government systems are being created on the local and regional levels. This development has potentially huge impacts on environmental protection activities, as environmental issues can be addressed in a more responsible and practical way because self-governments can directly assess the relative import-ance of different needs expressed by local populations. However, there is also a risk involved, as experience and expertise in resource management and environmental protection is lacking even more at the local level than at the central level. Furthermore, local communities tend also to have serious economic constraints and usually have fewer options than central governments for generating revenues (due to legal and other restrictions).

Market reforms and privatization   The CEE countries are attempting to restore market mechanisms based on private ownership and competitive trade. Successful transition to a well-established 'social market economy' is requiring a radical restructuring of the existing industrial systems, as well as improve-ments in technology and managerial practices. An influx of foreign capital, technology, and experiences continues to play an important role in catalysing the transition process. The cost of introducing a liberal market-orientated economy, after years of bureaucratic market control, is steep and being borne by all sectors of society.

Increased prices for energy and raw materials, as well as the gradual implementation of policies withdrawing government subsidies from industrial production, have created incentives for more efficient production practices and resulted in reduced pollution discharges. In Figures 1.6 and 1.7 the 1993 and 1994 prices for electricity and petrol in CEE countries are shown. Despite considerable progress since 1989, energy prices remain low compared to Western European levels, and generally remain below cost recovery levels. Prices for electricity are especially low, but other public services such as wastewater treatment (where available), water supply and waste management are also provided at prices far below actual costs (EBRD, 1995).

Related to achievements in price policy and subsidy policy, the decrease in physical output of heavy industry due to competition and market constraints, and the restructuring of industrial sectors, including the replacement of obsolete production technologies, have also contributed to pollution reduction. (See the country chapters for further analysis of these issues.)

The privatization of state owned enterprises is also a crucial process not only for increased efficiency and productivity, but also for setting free a spirit of entrepreneurship that can bring innovativeness and a new sense of responsibility. The privatization of production facilities should also go hand in hand with the resolution of past environmental liabilities. However, in practice, addressing this issue in a comprehensive manner has often turned out to be extremely difficult, as privatization is a complex and intergovernmental issue. Generally the problem has been addressed only on a case by case basis,

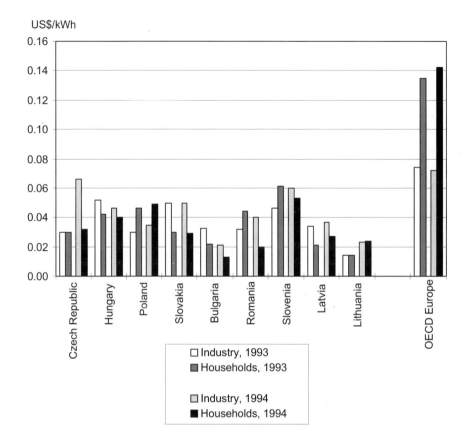

US$/kWh

Figure 1.6   Electricity prices for industrial and household use in CEE countries and
OECD Europe, 1993 and 1994 (US$/kWh). *Source*: UNECE (1995d)

though more comprehensive approaches are now being formulated in some
countries. However, in many cases the problem of environmental liability has
not been and is not addressed at all, or is addressed insufficiently (UNECE,
1995b), even though such neglected issues are likely to consume much time,
energy and money in the future.

The entrepreneurial spirit which has been set free in the region has led to
the rapid emergence of a great many small businesses. To this point such
development has often proceeded in a largely unregulated fashion. Quick
profits have become the dominant motivating factor among many businesses.
The 'profit now' attitude has become pervasive among much of the business
community and poses very real challenges for environmental protection in the
CEE countries. Competition is often intense, incentives to produce as cheaply
as possible are strong, and environmental protection expenditure is sometimes

US$/litre

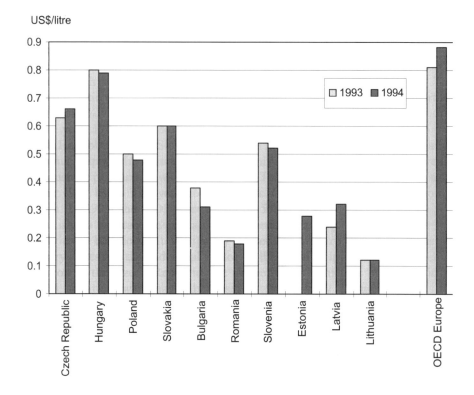

Figure 1.7   Prices of regular automotive fuel in CEE countries and OECD Europe, 1993 and 1994. *Note*: OECD Europe is based on sample of data available for Austria, Denmark and Germany. *Source*: UNECE (1995d)

forgone, with the costs of environmental degradation being externalized on to society. Moreover, short-term thinking and planning have become the norm, and not only in the private sector as politicians, government officials and citizens are also preoccupied with the immediate day-to-day stresses of the transformation and tend to ignore or undervalue the long-range effects of their actions (REC, 1994).

A successful market economy in today's world economic system must have a major international component. Businesses in the CEE countries are increasingly under pressure from foreign and domestic competitors. This can drive environmental protection in a positive direction (e.g. transfer of clean technologies and sound management practices), but outside pressure may also result in environmental degradation (e.g. waste imports, development of 'dirty' industries, wasteful consumption patterns).

Adoption of Western lifestyles   The growing desire of CEE citizens to attain Western living standards is providing the general motivation for production and growth. As free markets develop in the region, its citizens are fulfilling dramatically expanded roles as consumers. Consumers now have the possibility to purchase goods and services which they previously did not have access to. Advertising is playing a rapidly growing role in creating and shaping consumer demand and in establishing new consumption patterns. Environmental concerns, however, are rarely included in commercial advertisements in the CEE countries, reflecting the low environmental awareness of consumers and the priorities driving their decisions. Wasteful consumption is beginning to be an unfortunate symptom of development in the CEE countries, and the more modest (granted, there was little choice) and often environmentally responsible consumer behaviour of the past is disappearing.

Transportation practices are also evolving, generally in a most environment-unfriendly way. One example is the rapid growth in private passenger car use (see notes of Table 1.4), which has resulted in the existing transportation infrastructure being overwhelmed in many places, especially urban areas. The results of another change in consumer behaviour can be seen along streets and in wild garbage dumps throughout the region. The new waste stream, laden with plastics and proportionally less biodegradable material than traditionally produced in the region, is putting a major strain on the already underdeveloped and overburdened waste management systems of the countries. In many other ways as well, CEE citizens are adopting consumption patterns along the lines of unsustainable Western standards.

Social transformation   According to many accounts, in 1990 environmental protection was ranked as one of the top priorities of the public in most CEE countries. However, social and economic problems have since pushed environmental issues down on nearly all political agendas, and it has become obvious that solutions to environmental problems depend heavily on acceptable progress in economic development. This drop in concern for environmental issues was unintentionally helped by environmentalists who sometimes presented environmental programmes in isolation from the serious economic and social problems which confronted most citizens.

The average CEE citizen is facing serious economic difficulties (see Table 1.4 for a sampling of socio-economic indicators). Unemployment and job insecurity, falling real wages, and high inflation (all officially unknown in the past regimes) have pushed issues of economic welfare to the top of nearly all agendas. Pressing daily subsistence needs receive top priority and people are often not prepared to make judgements based on their long-term interests such as that of a healthy environment. Environmental preservation is simply not a high priority for many people, even though political changes have set the stage for citizens to be more involved in determining the future of their societies. Low public interest in environmental protection activities is also limiting the potential benefits of public participation, which should otherwise promote proactive pro-environment behaviour. The potential and importance of public

Table 1.4 Selected socio-economic indicators

Country	Life expectancy at birth[1] (years, 1994)	Unemployment[2] (%, end 1994)	Wages[2,a] (US$/month, 1993)	GNP/capita (1994)[1]		Urban population[1] (% of total pop., 1994)	Passenger cars[3,d] (cars/100 inh., 1990)
				Dollars[b]	PPPs, int'l $[c]		
Czech Rep.	73	3	221	3200	8900	65	21
Hungary	70	10	317	3840	6080	64	18
Poland	72	16	194	2410	5480	64	14
Slovak Rep.	72	15	201	2250	na	58	21
Bulgaria	71	13	114	1250	4380	70	14
Romania	70	11	82	1270	4090	55	6
Slovenia	74	15	421	7040	6230	63	na
Estonia	70	2	85	2820	4510	73	15
Latvia	68	7	98	2320	3220	73	11
Lithuania	69	4	65	1350	3290	71	13
Austria	77	...	...	24 630	19 560	55	39
France	78	...	...	23 420	19 670	73	42
UK	76	...	...	18 340	17 970	89	37
Portugal	75	...	...	9320	11 970	35	26

Sources: [1]The World Bank, 1996. [2]EBRD, 1994/5. [3]Eurostat, EC, EEA; 1995.

Notes: [a] Wages: Nominal monthly dollar wages at going exchange rates based on average 4th quarter 1993 earnings. [b] GNP/capita: Data are calculated using the World Bank Atlas method. [c] GNP/capita at PPPs: Purchase power parity (PPP) estimates of GNP/capita at current international dollars (extrapolated from 1993 ICP estimates). [d] Passenger cars: Data for Czech and Slovak Republics refer to the former CSFR. In CEE, there has been a strong increase in cars/capita since 1990: Available data indicate growth rates of 20–40% only in the period 1990–93. Such a dramatic increase occurred despite steep economic decline. Used cars from the West contribute with a high share of cars newly purchased in CEE countries since 1990.

participation itself is still not fully appreciated in the region, and possibilities for improvements in this area are not sufficiently taken advantage of (REC, 1994).

Moreover, mutually beneficial, positive relationships between social and environmental programmes have scarcely been examined and assessed in the CEE countries. The possibility of such relationships warrants concerted attention as many types of environmental protection efforts are labour intensive, and such efforts could provide 'win–win' solutions to both environmental and employment dilemmas.

While the former political systems of the region did not succeed in creating egalitarian societies, the social stratifications which did exist were probably not as extreme or conspicuous as those being created today in the CEE countries by the establishment of capitalism. Today's social strata are being defined according to source and level of personal income; a very visible wealthy upper class, and an equally visible poor lower class are developing. There are some signs that a strong middle class, typical of Western democracies, is evolving, but this process will take years and depend heavily on the rate and nature of economic progress.

The magnitude and conspicuous nature of such extremes in wealth, coupled with previously unexperienced social hardship, are inspiring sentiment to reintroduce extensive social welfare guarantees like those associated with the previous political system, evidenced by the return to power of 'post'-Communists in many countries of the region. The implementation of such guarantees, while perhaps defensible, can adversely affect environmental (and other) programmes as these guarantees require a diversion of funds from state budgets which are already under severe strain in the region. Cuts in environmental programmes have been suggested in some cases to balance economic and social requirements for financing (REC, 1994).

Economic Development Strategies and Environmental Policy

In the initial period after the fall of Communism, many environmental experts saw the dynamic changes about to come in the CEE countries not only as a unique opportunity to achieve immediate environmental improvements, but also as a way to firmly integrate environmental considerations into other policy areas *in the process* of the establishment of a new system. There were high expectations that the CEE countries could learn from the mistakes and successes of the West.

Today, however, it is clear that such expectations proved to be unrealistic. The idea that environmental protection and economic development are not exclusive, but rather mutually supportive is not widely held in CEE countries. It appears that most decision-makers in the region do not understand the significance of this basic integration of environmental and economic policies and therefore do not see the need for it. Unfortunately, environmentalists have been largely ineffective in changing this belief and have not been persuasive in demonstrating the fundamental link and possible harmony between economy and environment. While it is true that the environmental impact of particular decisions or programmes of economic policy is sometimes debated, little attention has been given to assessing the integrity of the overall transformation process and its likely longer-term environmental implications.

It sometimes appears that government and business decision-makers and environmentalists in CEE countries are locked in the environmental paradigm of the early 1970s where the apparent contradictions between economic development and environmental protection seemed to be fatal (cf. *The Limits to Growth* by D. Meadows et al., 1973). However, since *Our Common Future* (WCED, 1987) and the Rio de Janeiro 'Earth Summit' (UNCED, 1992), the idea of 'sustainable development' offers a new view most countries of the world (including the CEE countries) have endorsed, at least on paper. However, as in many Western countries, decision-makers in CEE have been reluctant to take the serious, far reaching steps necessary for implementing this new environmental paradigm.

Many observers share the opinion that the relatively frequent change of CEE governments and parties in political power, together with their often substantially differing political agendas, has led to additional difficulties in developing a cohesive transition in policy, economy, and social life. The situation has been made even more difficult by rather poor co-operation and co-ordination among governmental bodies which are themselves going through significant transformation (REC, 1994). Furthermore, the rapid rate of transformation has sometimes overwhelmed the capacities of these countries to handle the problems that arise from the process. To this point, environmental protection has not been a real factor influencing economic and financial decisions in the CEE region, and there is simply very little precedent in this region (or elsewhere it can be argued) for successful integration of environmental protection and economic development as priorities on an equal plane.

There are other impediments for linking environmental concerns strongly with economic policy. The basis for such a link has not been convincingly explained to the public. Thus, public demand for the inclusion of environmental concerns in economic development programmes remains weak. There are few legislative and policy mandates for such integration (despite support for 'sustainable development' being commonly expressed by government officials), and there is a lack of experience and expertise in planning and designing such laws and policies. Environmentalists often are not able (or willing) to understand the positions of businesses or government economists, while businesses and economists seldom place great value on the non-commercial use of environmental resources (REC, 1994). Effective integration of environmental protection and economic development will require a new type of thinking in CEE countries.

Observations on selected achievements, problems and trends in environmental protection in CEE countries

Individual and Institutional Capacity

Resolution of the tremendous environmental challenges facing CEE countries will require impressive human effort, on the individual and institutional levels. In order to meet these challenges individuals and institutions in CEE countries must continue to improve and expand their capacities in a broad range of areas. The Communist systems of the past often neglected, rejected or just had no use for many of the kinds of skills and abilities required to design and implement effective environmental protection systems. Decisions having significant environmental (and other) impacts were too often based on political doctrine rather than sound science or economics. Not only must the individuals and institutions now involved in environmental protection address weaknesses resulting from the past, they must also adapt themselves to the dramatically different and changing circumstances brought on with the advent of democracy and market economies. In the current CEE setting, where scarce resources are inadequate to meet existing needs and citizens demand responsive governments, it is not enough for environmental protection efforts to merely be effective, they must also be efficient and should be understood and accepted by the public. This new reality drives the need to develop new skills and capacities as well as better understanding and greater appreciation of environmental protection issues.

In the CEE countries fundamentally new policies, procedures and institutions are being created. Currently in most (if not all) CEE countries the relevant government administrative and political decision-making bodies are the key actors. This is true for all sectors, including environment protection. However, government environmental protection authorities are often unable to carry out their mandates sufficiently for a number of reasons, not the least of which is the relatively very weak positions they hold in comparison to other sectoral institutions. Throughout CEE, the status and power of environmental ministries is much lower than most other government ministries and efforts of the environmental authorities are frequently rejected, over-ridden or simply ignored by the more powerful government sectors (i.e. Ministries of Finance, Economic Development, Industry). This can be clearly illustrated by the many cases where official policies of other government sectors are in direct contradiction with approved state environmental policies. (A perfect example is in Poland, where the government officially adopted 'sustainable development' as the guiding principle for economic development and environmental protection in 1991, but recently approved a scheme for building a huge new road and highway network throughout the country, as the existing railway network shrinks and deteriorates and railway officials scramble for funds.)

Fundamental shortcomings in the areas of environmental management and policy-making within environmental authorities, independent organizations and even private sector enterprises hinder progress in the development of pragmatic, cohesive, and effective decision-making and planning processes. Such shortcomings lead to a host of problems, such as: poorly defined and ambiguously allocated responsibilities within and between institutions; the inability to set clear priorities; insufficient capacity to design comprehensive yet feasible implementation plans; inadequate budgetary planning and management; weak monitoring and evaluation schemes, etc. While none of these problems is solely environmental in nature, their resolution bears heavily on the success of any environmental protection efforts. Western experience and expertise in the areas of environmental management and policy-making, though often not directly transferable to the Eastern setting, can be useful in helping CEE environmental specialists and institutions to meet the new and growing challenges facing them. In addition, strong and enlightened leadership in government administrations, parliaments, and the private and independent sectors is essential.

Besides better skilled managers, successful environmental programmes require qualified staff who are capable of meeting the demands of existing challenges and contemporary conditions. Unfortunately, there is a lack of such staff people in CEE institutions at this time. Extensive training is required to address this need and, indeed, many environmental training programmes of a wide variety have been offered in recent years and many more are underway or planned. When qualified staff do exist in government administrations, however, they are difficult to retain because of the low salary levels. While there are many qualified people and scientific specialists in the region, this expertise tends to be narrowly focused or technical in nature. Todays circumstances require that technical experts be complemented by people who have strong managerial skills and are able to communicate effectively with decision-makers and the public.

Standard business practices also strongly influence the effectiveness of environmental protection endeavours. As in government organizations of the region, environmental management capacities in business are far from ideal. Though economic reforms have forced businesses to become more conscious about their use of natural resources and production of waste by-products, commercial enterprises still lack experience in responsible environmental management and their willingness to pay for environmental protection is often low. The situation is exacerbated by the competitive pressures businesses are under to cut costs wherever possible and boost profits.

Enforceable Environmental Policy and Law

In all CEE countries, broad environmental policies or strategies on the national level have been or are being developed and put into force. In some countries environmental strategies have also been developed on the regional or local level. (More in-depth analysis of environmental policy development in the individual countries can be found in the country chapters.) In spite of important achievements in the development of environmental strategies, major problems remain. Taking into account all the adverse circumstances affecting environmental protection efforts in the region, it is not surprising that such environmental programmes often represent more wishful thinking than feasible action plans. The most fundamental challenge in most CEE countries is to translate environmental policy goals into coherent, action-orientated programmes, which can actually bring about the desired improvements. An environmental policy can only be effective if the capacity and will to implement such a policy exists, which is not always the case in CEE countries. Furthermore, the obstacles impeding implementation of environmental programmes are often not sufficiently evaluated or anticipated. (The most important problems related to developing effective environmental policies in CEE countries are reflected in the EAP–NEAP process, discussed below in the section on the environment for Europe.)

The state of environmental law varies considerably among the countries of the region. While some have developed quite extensive contemporary legislation, others are still relying on old laws, established under previous political systems, which are largely inappropriate for the current circumstances. Some gaps and weaknesses in environmental law exist in all of the countries, however, modern framework legislation for environmental protection has at least been proposed in all of the countries. Typically, legislation enacted after 1989 has established key policy principles (such as the 'polluter pays' principle, access to environmental information and public participation), provided new policy tools (such as pollution charges and taxes, and Environmental Impact Assessment) and established new environmental institutions (such as environmental funds or monitoring and enforcement agencies) (modified after EAP Task Force, 1995b). More detailed information on the state of environmental law in the individual countries can be found in the country chapters.

There remains a strong need for environmental law in the region to be made more comprehensive and more practical. 'Practical' environmental law is emphasized, on the one hand, because currently law enforcement is regarded as the weakest element of existing environmental protection systems in the CEE countries. On the other hand, law which is too strict renders enforcement unrealistic and provides incentives for illegal activities. It should be noted that the fundamental principle of rule of law is not yet as firmly established in CEE countries as it is typically in most Western countries and years will be required for the necessary institutional framework and societal attitudes to be formed which can allow such development. Most important for the development of legislation, regulations and policy in the foreseeable future will be European Union law and policies. (For deeper discussion of issues related to EU accession please see the following section.) Table 1.5 gives an indication of the progress CEE countries have made in harmonizing their environmental protection legislation with EU law.

Table 1.5 Estimated approximation levels of CEE countries with European Union environmental legislation (in %, as of autumn 1995)

	BUL	CZE	EST	HUN	LAT	LIT	POL	ROM	SR	SLO
General environmental policy regulation	55	77	33	87	22	67	63	38	72	55
Air	43	60	40	40	23	40	47	37	77	50
Chemicals, industrial risks, biotechnology	27	31	35	14	19	19	33	32	33	23
Nature conservation	67	67	33	67	50	67	100	33	100	67
Noise	50	17	17	50	17	0	50	17	50	50
Waste	24	56	45	18	14	24	26	21	78	19
Water	50	61	72	61	44	56	78	61	78	44

Source: Regional Environmental Centre, 1996.

Note: The figures above are rough estimates made by environmental law experts from the respective countries and provide an indication of the situation at end-1995. The figures provide an indication of the level of harmonization of written law, and do not necessarily indicate the degree of actual compliance in the CEE countries. BUL = Bulgaria, CZE = Czech Republic, EST = Estonia, HUN = Hungary, LAT = Latvia, LIT = Lithuania, POL = Poland, ROM = Romania, SR = Slovak Republic, SLO = Slovenia.

Public Participation and Information

Environmental coalitions and interest groups continue to evolve and mature in the CEE region. It seems likely that this evolution process will be similar to Western experience. As the process of professionalization and institutionalization progresses within the newly-established structures, stronger lobbies, pro- and anti-environment, can be expected to arise.

There are already a myriad of independent environmental groups in the CEE region (about 1800), some of which are well organized and trained (REC, 1994). However, for a number of reasons, including the following, NGOs in CEE have not realized their full potential: funding constraints; poor co-operation; lack of management, planning and other skills; and sometimes an unwillingness to compromise. With few exceptions, NGOs have had influence on national environmental policy development. Most of their activities and successes have taken place at the local level and on very specific issues.

A feature of Communist rule in CEE countries was that comprehensive environmental information was not collected, and what was collected was not made available for public use. When environmental data and information were presented, the accuracy was generally questionable (to put it mildly). This problem is still relevant for policy-making today. For instance, there remain major difficulties in compiling accurate historical data lines of emissions and environmental conditions.

With the collapse of Communism, access to, and availability of, environmental information has improved. For example, state of the environment reports have since been published in most CEE countries, allowing for improved assessment of damages. However, it is still difficult to compare the quality and thoroughness of such reports with that of similar reports prepared in Western countries.

Even if the amount and quality of information is continuously increasing, and the very basic information necessary for action-orientated decision-making is available, major challenges persist. Ministries of Environment and other relevant governmental bodies are often criticized for making available far too little, and simplistic or incomplete information (REC, 1994). Indeed, better information policies would contribute to stronger public discussion and support of environmental issues and should be a priority for environmental authorities throughout the region.

Technological Resources

Related to the historical development of industry and the lack of funds in the region is the need for modern technologies. Many of the environmental problems in the CEE countries result from the operation of obsolete, inefficient production systems using outdated technology in conjunction with poorly functioning or completely absent environmental protection equipment.

New production and environmental technologies of all types are needed but they are slow in coming as the costs are often prohibitive for the public as well as private sectors. Management measures that improve the use of natural resources and promote waste minimization and recycling practices are also very much needed. Pollution prevention and waste minimization are not standard business practices in the CEE countries, and here East–West co-operation can be crucial in catalysing improvements. A number of bilateral and multilateral technical assistance programmes are in progress throughout the region and an important goal will be to make such activities sustainable after the foreign assistance ends.

Financing Environmental Improvements

A considerable variety of institutions and mechanisms are involved in the financing of environmental protection in CEE countries. (Some of these are discussed in more detail in this chapter's later section on environmental finance.) Generally, these institutions and mechanisms should be seen as tools or instruments supporting the implementation of state (or sub-national) environmental policies. The institutions include those typically involved in environmental protection – the state Ministries of Environment, Environmental Inspectorates, regional and local environmental authorities – as well as specialized financial institutions such as environmental funds and banks. Of course, polluting enterprises themselves are also involved in financing environmental improvements and will become increasingly so as the 'polluter pays' principle comes more into practice. These different institutions are responsible for creating, implementing or complying with the various types of mechanisms and instruments used in financing environmental protection in CEE countries.

Ministries of Environment have created or initiated economic instruments such as pollution charges and fines, environmental taxes and product charges, tax reliefs, even tradable emission permit systems (on a pilot project basis in Poland and in draft legislation in Latvia). These instruments generally serve two functions: (i) they provide incentives for polluting entities to eliminate or restrict their polluting behaviour or, alternatively, they promote environment-friendly behaviour; and (ii) they raise revenues which are often used for environmental protection purposes. In practice, however, the revenue raising function of economic instruments is much more important than the incentive function (Klarer, 1994), and there are only a few exceptions such as the Polish system of environmental charges. As can be seen from the following country chapters, some of the CEE countries have developed fairly extensive systems for financing environmental improvements, while others are still in the early stages of devising such systems.

The establishment of environmental funds by many CEE countries illustrates the urgent need for a stable flow of funds for environmental protection measures. However, the necessity of using such extra-budgetary mechanisms also demonstrates a lack of integrity in the overall economic and environmental programmes. With the creation and use of the funds one may reasonably conclude that the CEE governments have deemed separate programmes for the support of strictly environmental spending to be a priority, while a more integrated approach simultaneously supporting environment-friendly and beneficial investment and economic development is hardly to be found anywhere in the region on more than a local or demonstration project basis, despite the endorsements of 'sustainable development' commonly heard from officials throughout the region.

The role of public spending for environmental protection will remain significant in the transition period. While market forces are playing a growing role in encouraging environmental investments in the private sector, governments will continue to provide significant financing for environmental improvements for some time to come. There is still a large state-owned sector of the economy that requires environmental subsidies and the State must also continue to provide funds for local and regional governments to meet their environmental requirements, even though decentralization is the trend in most CEE countries.

CEE Regional Co-operation on Environmental Issues

The transition to democracy and a market economy is a task which the CEE countries are compelled to face largely on their own. The CEE countries have a common environmental heritage due to the (il)-logic of the former system and their present and future development paths will also share much in common and will even be interrelated. However, regional co-operation has only minimally been realized. There are few examples of co-operation and support between the more advanced CEE countries and their struggling neighbours and meaningful co-operation between the more advanced countries is also lacking. This does not help the overall transformation and could lead to further fragmentation of the CEE region.

There are several factors hindering environmental co-operation among the CEE countries. A lack of political will for regional co-operation is evident. Some of the motivations behind this stem from rivalry and competition to be the first country in the region to be integrated into the European Union. The chances for integration would increase if the CEE countries could demonstrate their capability of handling their own problems. Co-operation in the CEE region has also been limited by the countries' disinterest in dealing with other countries who cannot financially assist them. There is also a kind of backlash against co-operation within the region resulting from the many years of forced 'co-operation' and involuntary 'brotherhood' imposed by the Communist regime.

Competition for Western environmental financial assistance also sometimes reduces the incentive for co-operation within the CEE region. Such competition is all the more regrettable as the Western solutions offered for tackling environmental problems in the CEE countries are not always adaptable or the most appropriate. In some cases, CEE countries could provide more effective support to each other through transferring positive experiences or solutions which have been developed and tested in the region.

Environment Related Foreign Assistance

A considerable amount of foreign environmental aid has flowed into the region in recent years. Numerous examples of successful environmental protection activities which were supported by foreign donors could be cited: training has been provided to government officials, business managers and NGO activists; environmental policies and institutional structures have been strengthened; environmental assessments, inventories and feasibility studies have been conducted; monitoring systems have been designed and established; and major investments in technological capital have been made. Early assistance was usually in the form of feasibility studies, investigations, institutional strengthening and technology transfers and advice for solving the most acute problems.

Environmental assistance has also helped to encourage public participation in environmental protection projects and strengthened local implementation capacities. In some cases, there have been formal requests from donors to use participatory procedures. Steering committees have also been used as a tool for involving local people in projects. This type of development can have a tremendous impact on establishing democratic procedures in different decision-making bodies and economic sectors. In the course of implementing projects supported by foreign assistance, it has become clear that the engagement of CEE national and local experts is critical for achieving positive results, despite their sometimes lacking management and strategic planning skills. Through the process of participating in foreign aid supported activities, project co-ordination skills have been acquired by many CEE government and business managers.

On the other hand, there have been a wide range of negative experiences with foreign assistance as well. A large amount of money has been invested in assessments and feasibility studies, often without resulting in practical improvements in the environment. Moreover, these numerous studies and reports have often not been prepared in cost-efficient ways, using very costly Western consultants who lacked important local knowledge and made only short-term visits to the region. These studies have often failed to meaningfully involve local and national experts who have direct knowledge of the issues. In other cases, the priorities of the donor organizations are simply not compatible with the priorities of the recipient CEE country.

There is a strong preference among donors and CEE countries to develop bilateral relations. This can be an obstacle to undertaking regional environmental programmes and to encouraging regional environmental co-operation. Western interest is also gradually shifting from the CEE region towards the Newly Independent States (NIS). Perhaps considered a more important, more strategic future political partner, the NIS may receive assistance at the expense of the CEE region. It must also be said that the needs for assistance in some CEE countries are steadily decreasing and are much lower than in many NIS countries. This is illustrated by the fact that three CEE countries, the Czech Republic, Hungary and Poland, have already joined the OECD (known as the club for the world's wealthiest nations), and the Slovak Republic is soon to follow.

In many cases, foreign aid is playing the role of catalyst in environmental improvements. Withdrawal of Western technical and financial assistance from the CEE countries would reduce incentives and an important portion of financial means which support environmental improvements in the region. The process of modernizing environmental laws, designing market based protection mechanisms and formulating improved policies would also be adversely affected. For some countries of the region, Western assistance is essential for acquiring new skills and technology. Maintenance of future foreign assistance for environmental improvements is also important for encouraging and assisting co-operation within the CEE countries and to aid serious consideration of sustainable development strategies. Foreign financial assistance is very limited and should, for the most part, focus on the most immediate and serious environmental threats, while supporting the development of policies and mechanisms capable of achieving long-term benefits.

1.2 THE EUROPEAN DIMENSION: TOWARDS CONVERGENCE OF ENVIRONMENTAL POLICIES

The introduction of democratic political systems, market economies, and the opening of borders, has established the foundation for CEE countries to achieve full participation in international development and co-operation. International co-operation has become increasingly important in the recent years to tackle global and regional problems, as well as to strengthen national environmental policies.

In this section, the discussion focuses on issues related to environmental policy at the European level and the process of convergence of environmental policies in Europe, as well as the role and significance of CEE countries within this process. It should be noted that global issues, nor environmental conventions and agreements or other such forms of international co-operation are discussed here, though they certainly are important and can have major effects on national environmental policies of CEE countries.

The Environment for Europe process

Evolution and Present State of the Environment for Europe Process

The 'Environment for Europe process' refers to environmental policy development and convergence at the European level, centred around the Pan-European Environment Ministers' conferences held in 1991 at Dobris in the former Czechoslovakia, in 1993 at Lucerne, Switzerland, and in 1995 at Sofia, Bulgaria. (The next Pan European Ministers' Conference on the Environment for Europe is planned to take place in Denmark in 1998.) The special significance of the Environment for Europe process is obvious if its broad and comprehensive approach is considered with the involvement of OECD, Western European countries, CEE and NIS countries, the European Union, the major international finance institutions (IFIs), as well as a number of environmental non-governmental organizations (NGOs) and other international organizations. The Environment for Europe process presents a formidable platform for contact and co-operation between East and West with the representation of all mentioned countries and organizations. The Environment for Europe process is unique, no similarly comprehensive effort exists in other policy areas.

In light of the changes in CEE and growing evidence about the true extent of environmental problems there, and following the 1990 Bergen and Dublin conferences, the first Pan-European Environmental Ministers' Conference, including the CEE countries' representatives, took place in 1991 in the Dobris castle of former Czechoslovakia. At the Dobris conference, ministers agreed to strengthen co-operation on the development of European environmental policy and to give special attention to CEE problems. This special attention has been maintained until today. Since 1993, the successor states of the former Soviet Union (Newly Independent States – NIS) are increasingly included and covered in the relevant work of the Environment for Europe process. The Dobris conference also called for a report on the state of the pan-European environment and an improvement of the situation with respect to environmental policy-related information in CEE. Following the Dobris conference, four task forces were formed.

The first task force, led by the World Bank, the OECD Environment Directorate and the European Commission Directorate General XI, was established for preparing an environmental action programme for CEE. The aim was to prepare a strategy to guide immediate action and to tackle the most grave and urgent environmental problems of CEE, especially where human health is endangered, where a risk of irreversible damage to nature exists, and where serious economic losses were caused by environmental degradation. The work on such a strategy was backed by the preparation of a number of reports addressing specific policy areas. (Among other issues, special reports covering the CEE area were prepared on: environment and health; privatization and environmental liability; foreign direct investment and environment; financing and environment; and sectoral studies – environment and energy/industry/agriculture.) The result of these efforts, the Environmental Action Programme for Central and Eastern Europe (EAP), was adopted at the 1993 Lucerne conference. The EAP is discussed in more detail below.

A second task force was formed, led by the European Commission, to prepare a comprehensive report on the state of the European environment. The mandate was carried out by the European Environment Agency (EEA). The result of this huge effort, the first state of the environment report covering the whole of Europe including CEE (see Europe's Environment – The Dobris Assessment, 1995, and, Eurostat, 1995), was finished in 1995 and was presented to the 1995 Sofia Ministerial conference. Simultaneously, a third task force led by the United Nations Economic Commission for Europe (UNECE) received the mandate to prepare an Environmental Programme for Europe[2] building on the relevant results of the European state of the environment report. The Environmental Programme for Europe was endorsed by the Sofia Ministerial conference.

The fourth task force, led by the European Council and the International Union for Nature Conservation (IUCN), developed the Pan-European Biological and Landscape Diversity Strategy, which was also endorsed by the Ministers' conference in Sofia. The adoption of this strategy and future activities related to it, will be of major importance for CEE countries, since large parts of the region remain relatively untouched by humans. The protection of such areas must be ensured, especially in the context of ongoing and future land privatization and rural development strategies. Another task force will be created, led by the Council of Europe and UNEP, in co-operation with OECD and IUCN, to guide and co-ordinate the implementation and the further development of the strategy (Sofia Declaration, 1995).

There were clear signs from the recent Ministerial conference in Sofia (October 1995), that East–East co-operation should be strengthened and that CEE countries should take over more ownership and responsibility in the Environment for Europe process. It has also become obvious, that stronger public support must be sought to further increase the success of the process. The Sofia Ministerial conference received a relatively low level of media and public attention both in Western and CEE countries. This can partially be explained by the complexity of the process and subject matter involved. Indeed, there was rather strong, wide-ranging criticism that the Environment for Europe process should be simplified in the future.

The Sofia Ministerial conference also stressed that more account needs to be taken in the Environment for Europe process of the European Union accession process for CEE countries with Europe Agreements (see also the section on European Union accession below). It was suggested that the relations between the Environment for Europe process and global processes (such as the Rio process/Agenda 21), regional processes (e.g. the Black Sea or

Danube environmental programmes), as well as other programmes on the international levels ('Environment and Health', 'Environment and Transportation'), must be re-considered and synergistic effects amplified.

It should be mentioned that prior to the Sofia Ministerial conference, CEE Environment Ministers met in a consultation and identified six areas where progress was particularly encouraging, and which should receive additional attention to help accelerate the further development of environmental protection in CEE countries. These areas, also called 'the Sofia Initiatives', were: (i) environmental impact assessment; (ii) economic instruments; (iii) project preparation capacity; (iv) reduction in toxic air pollution (heavy metals), including promotion of lead-free fuel; (v) reduction in SO_x emissions; and (vi) biodiversity (EAP Task Force, 1995b). The 'Sofia Initiatives' were welcomed and supported by the Sofia Ministerial conference.

Today there are two main bodies responsible for carrying out the mandates and work adopted at the Ministerial conferences in Lucerne and Sofia: The Task Force for the Implementation of the Environmental Action Programme for CEE ('EAP Task Force', with the secretariat at OECD, Paris); and the Project Preparation Committee ('PPC', with the Secretariat at EBRD, London). In addition, the Working Group of Senior Governmental Officials (with its secretariat at UNECE, Geneva) is the main co-ordinating organ of the Environment for Europe process.

The PPC has the role of co-ordinating and facilitating donor and IFI involvement in environmental investments in CEE and is discussed further below in the section on financing environmental protection in CEE countries).

The goals of the EAP Task Force are to assist CEE countries to identify and tackle priority environmental problems in the most cost-effective manner and to establish a better basis for dialogue and co-operation with donors. The needs and priorities of CEE countries provide the basis for the work of the EAP Task Force, and measures along the lines recommended in the EAP provide the substance of its activities. In its work, the EAP Task Force provides a framework for know-how exchange and informal co-operation among CEE countries on a broad range of policy and institutional issues (EAP Task Force, 1995b). The three major areas of activity in the framework of the present EAP Task Force Secretariat work are: 'EAP–NEAP'; 'financing'; and 'business, industry and environment'. These three areas are further discussed below.

From the EAP to NEAPs

The Environmental Action Programme for Central and Eastern Europe (EAP) was developed to help address environmental problems in CEE countries that require urgent action. It was intended that this action should be taken in the 'window of opportunity' which followed the collapse of the old regimes in order to remedy past pollution as rapidly and cheaply as possible in the context of economic transformation, and to prevent new pollution problems from arising in the future by establishing new patterns of environmental management. Environmental health related issues were identified as a priority.

The EAP did not set environmental quality targets nor did it estimate costs of reaching such targets. Its goal was to help ensure that the most urgent environmental problems of CEE could be solved as cost-effectively and efficiently as possible, by promoting the use of a mix of policy instruments that include market-based solutions, realistic regulations and enforcement mechanisms, and carefully selected and prepared investments.

The EAP emphasized the strong linkages between institutional capacity building, appropriate policy frameworks and the preparation of investments to reduce emissions. These three areas represent the fundamental pillars upon which the EAP was built. In Box 1.1, the principal recommendations made in the EAP are presented. In addition to these key messages, the EAP also recommended a number of very practical actions and measures which should receive priority attention from environmental policy decision-makers in CEE countries.

The significance of the EAP in practice is that it was based on a multi-country initiative (with heavy involvement of the World Bank in the preparation process) and adopted by all European countries. The EAP is intended to serve as a strategic steering document for the most important environmental investment programmes in the region. In addition, the EAP offers basic guidance for the provision of Western technical assistance to the region and the development of environmental policy in CEE countries.

Especially after adoption of the EAP in 1993, a number of unrealistic expectations and misunderstandings of the EAP's role hindered quick and practical implementation of its maintenants. Some CEE voices criticized the EAP as a Western document. Critics felt that they did not need to be told what needs to be done in their country, especially if no substantial investments from the West would be connected with implementation of the EAP. Another criticism was that Western experts operating in CEE often do have a tendency to dominate joint discussion and decision-making.

A further impediment to implementation was that the EAP was available for some time only in the English language, and therefore broad dissemination of the document in CEE was not possible. The translation of the EAP in 1994–95 into all languages of CEE signatory states must be seen as an important step for enabling the adoption and implementation of its main recommendations.

Successful implementation of those recommendations, however, require that they be carefully tailored into the context of each country's specific environmental problems and transitional circumstances. To enable this, the process of preparing National Environmental Action Plans (NEAPs), based on the principles of the EAP, is dominating the EAP implementation efforts.

The NEAP concept shares many aims of the EAP, including the aims to reduce environmental pollution and safeguard natural resources at lowest cost

Box 1.1   Principal recommendations of the Environmental Action Programme (EAP)

– Countries have many opportunities to implement policies and *invest in projects which provide both economic and environmental benefits.* 'Win–win' policies include removing subsidies that encourage the excessive use of fossil fuels and water in industry, agriculture and households. They also include investments in energy and water conservation, low-input and low-waste technologies, and expenditures on 'good industrial housekeeping'.

– Environmental priorities should be based on a *careful comparison of costs and benefits.* The resources available for environmental improvements will be severely constrained in CEE for at least the next 5–10 years. It is essential that limited resources be applied to the most urgent problems first.

– *Market forces should be harnessed to control pollution* wherever possible. Market-based instruments, such as pollution charges, fuel taxes, and deposit refund schemes, can help achieve desired levels of environmental quality at much lower costs than traditional regulatory approaches. Regulatory instruments will still be needed to control emissions of micro-pollutants such as heavy metals – particularly lead – and toxic chemicals.

– Countries should *concentrate on local problems first.* Many people suffer health damage from exposure to lead in air and soil, airborne dust and sulphur dioxide, from nitrates in drinking water and from contaminants in water and food. Solving these problems will bring the biggest gains in health and well-being. Measures to reduce emissions of pollutants in response to local concerns should also contribute to reducing transboundary and global emissions.

– *Standards need to be realistic and enforceable.* Countries should implement stricter standards over a 10–20 year period, and ensure that industries comply with interim standards. Local people should be involved in setting priorities and in implementing solutions. Neither governments nor donor institutions are equipped to judge how local inhabitants value their environment. A *participatory approach* is essential for the long-run sustainability of environmental improvements.

– *Responsibility for past environmental damage needs to be clarified.* Uncertainty about who will be responsible for past damage can discourage foreign and domestic investment and can impede privatization. For practical reasons, governments will have to bear most of the costs of dealing with past pollution. Governments must define clearly the environmental standards that new owners must meet and the period of adjustment that will be permitted.

– Donor countries should consider providing funding to *accelerate the reduction of transboundary and global emissions* in countries of CEE. Such funding would be particularly appropriate where the marginal costs of reducing emissions is lower in Central and Eastern Europe. Minimizing the net cost of meeting international agreements is in the interests of individual countries and Europe as a whole. If the net cost of reducing transboundary flows is lowered, countries will be able to afford to act earlier or to adopt more stringent targets.

– More *research, training and exchange of information* are needed to help decision-makers set sensible priorities. Research should focus on the state of the particular environment of CEE. Much more information is also needed on low-cost ways to reduce emissions of air and water pollutants from non-ferrous metal smelters, iron and steel plants, chemical plants, paper mills and waste water treatment plants and on ways to conserve biodiversity.

– Finding, implementing and financing solutions will require *building partnerships.* Transferring know-how and clean technologies will require strong co-operation between East and West, between countries of Eastern and Central Europe, and within the countries, between cities, institutions, and enterprises.

Source: World Bank, 1994.

as quickly and effectively as possible, to build environmental management capacity, and to improve communication on environmental problems and domestic solutions. NEAPs are understood to be part of a longer term and goal-driven process of capacity development. Perhaps the most important feature of the NEAP concept is that it must be 'locally owned', i.e. it should be prepared and implemented by experts in the respective countries.

The EAP Task Force recommends that NEAPs be built on the following key structural elements (modified after EAP Task Force, 1995d):

– Clearly defined criteria for choosing key sectors and problems.
– Targets toward progress can be measured.
– Ranking of problems in order of importance.
– Analysis supported by relevant environmental, economic and other information (e.g. cost–benefit analysis).
– Realistic budget estimates.
– Investment plans geared to available resources (e.g. from environment funds, foreign assistance).
– Implementation plans with timetables.
– Methods for communicating/negotiating with those affected by the plan ('stakeholders').
– An accompanying monitoring and evaluation process.

Keeping in mind the various problems related to environmental policy discussed earlier in this chapter, it is obvious from the above list that the development of effective NEAPs will be a difficult task and will require many years. Only one major issue will be mentioned here: priority setting. Ranking problems and sectors in order of importance (e.g. according to health criteria) and ranking remediation actions according to cost-effectiveness is impeded by a lack of consistent and comparable data. Data on health effects are frequently weak (as in the West) and there is a general lack of capacity for economic analysis in all CEE countries. In addition, priority setting requires an ever elusive political consensus: political will and motivation are crucial for the development of a successful NEAP.

Despite the obstacles, the development of NEAPs in the region is very promising for several reasons. It can be expected that the capacity to prepare pragmatic and feasible programmes will be enhanced in a learning-by-doing process. In addition, as the whole effort is taking place in the framework of the Environment for Europe process, strong emphasis is placed on experience exchange between CEE and Western countries. This exchange is actively encouraged and supported by a wide range of indigenous institutions and advice and training are being provided by many Western sponsored technical assistance programmes.

The process of preparing documents called NEAPs started in 1994. Today, Ministries of Environment are working on NEAPs in all CEE countries.

Experience shows that three main strategies for developing NEAPs are being pursued (EAP Task Force, 1995d). Some CEE countries already have broad environmental strategies. In this case, a NEAP might start from the comprehensive policy goals, and aim to implement one or more of their most important elements. The NEAP will be more limited in scope and have a shorter time horizon than the broad strategy document. This is the process already developing in Slovakia, Romania and Poland.

In countries where no broad environmental strategy was previously prepared, new strategies and action-oriented NEAPs are being formulated jointly (as in Estonia, Latvia and Lithuania). In these cases, more attention has been paid to drawing together available environmental information, consulting the public and developing closer communication with other Ministries. In Slovenia, preparation of a comprehensive strategy and of a NEAP have been parallel processes, generating much discussion both within the Ministry of Environment and with relevant sectoral Ministries.

A third strategy is found where an external consultant's report has been adopted by the country. Country strategies prepared by the World Bank in conjunction with national experts and Ministry of Environment representatives have been accepted by Albania and Bulgaria as National Environmental Action Plans to be put into effect. These strategies have a mix of regulatory, economic and investment elements. Experience has shown that they form a useful checklist of actions concerning the regulatory and economic instruments sectors. But the necessary investments have been slow to materialize.

Financing

Huge needs for investments exist in practically all sectors of CEE countries, including, and perhaps especially, the environmental sector given its neglect in the past and subsequent deterioration. While the financing of environmental protection in CEE countries is discussed in greater detail in a later section of this chapter, the importance of the subject within the Environment for Europe process warrants a brief introduction here. The later discussion will illustrate and examine some of the main obstacles to financing environmental protection in CEE countries as well as major sources of environmental finance in the region and emerging finance mechanisms. The EAP Task Force and its investment-oriented counterpart – the Project Preparation Committee – are working to increase understanding of the different mechanisms and institutions involved in financing environmental protection in the region and the associated problems. (The Project Preparation Committee was initiated within the Environment for Europe process to focus on the third main element of the EAP – implementation of priority environmental investments.) As will be seen in the later review, financing and investment related activities are receiving growing emphasis within overall EAP efforts as they are critical to capturing the benefits to be gained from reforms in environmental policy and institutional strengthening and necessary for the realization of real environmental improvements.

Business, Industry and Environment

In mid-1995, the private sector share of GDP was 40–70% in CEE countries. The fastest growing private sectors are services, agriculture and food processing, which are diffuse sources of pollution. In most CEE countries the major polluting industries remain state owned, many of which are likely to stay in business for the medium term (EAP Task Force, 1995b). Foreign and domestic direct investment in industry and other enterprises is steadily growing in CEE countries, especially in countries advanced in economic transition and close to Western Europe. Progress in European Union accession and alignment of regulations and practices with respect to the internal market of the EU will provide conditions inducing a further boost of overall direct investment.

Trade liberalization has been accompanied by the collapse of much intra-CEE trade. Trade with OECD countries, particularly Western European countries, has expanded correspondingly. This, in turn, has raised questions about limitations on market access related to environmental requirements in the EU Member States. Some argue that compliance with product standards, eco-labelling and other requirements in the EU is providing one of the most powerful incentives for improved environmental performance in the export-oriented sectors of CEE industry (EAP Task Force, 1995b).

Foreign direct investment is also transferring cleaner technologies and environmental management know-how into the region. Increasingly, CEE countries are challenged to develop balanced and equitable strategies for addressing priority environmental problems in both the public and private sector. Together with an appropriate framework of incentives and regulations, cleaner production programmes can be a particularly effective means for improving the environmental performance of the industrial sector. Results obtained in CEE in co-operative programmes (UNECE, 1995c) involving Norway, the US, Denmark and others indicate that 20–40% reductions in wastes and emissions have been possible for little investment. A further 30% reduction is possible through investment in technically proven and profitable equipment or process change.

For Western companies evaluating specific investment opportunities in CEE countries, environmental factors can be an important element in the decision whether to proceed. One factor might be past environmental contamination on a site, and another whether compliance with the host country's environmental standards is possible. In the last four years, privatization and environment officials have gained much experience in ways and means to consider environmental liability issues in the privatization process. However, many important challenges remain, especially with respect to the fact that most heavy industries have not yet been privatized.

In recent years there has been a flurry of activities related to business, industry and environment and the EAP Task Force has initiated a work programme in this area. During a number of conferences, workshops and meetings the following key issues have been identified for further work:

- Harmonization of environmental regulatory standards with those of the EU and OECD countries (e.g. environmental management standards, environmental audits, reporting, etc.).
- Harmonization of environmental standards and implications for trade.
- Privatization, foreign investment and environmental liabilities.
- Establishing toxic release inventories or similar mechanisms.
- Environmental management and clean production programmes/centres.
- Building environmental capacity in business (education and training), especially for small and medium-sized enterprises.
- Establishing regulatory frameworks which can provide appropriate incentives for action, but also allow voluntary commitments by business/ industry.

Various international and national organizations are already active in the above areas. Some of these are the World Business Council for Sustainable Development (WBCSD), the International Chamber of Commerce (ICC), the International Network for Environmental Management (INEM), Environmental Management Associations (in the Czech Republic, Hungary, and Slovenia), and the Bulgarian Industrial Association. By strongly involving these and other organizations in dialogue about business, industry and environment, the developing private sector in CEE countries is expected to be more directly engaged in the Environment for Europe process.

All major Western assistance programmes, bilateral and multilateral, are able to participate in future EAP Task Force work related to Business, Industry and Environment. There are good chances that through these assistance channels, better management and low cost investments in the worst polluting industries can be encouraged, cleaner production programmes can be assessed and broadened, and further exchange of experience between Western and CEE companies can be encouraged. The EAP Task Force also intends to develop best practices guidelines for environmental management and disseminate case studies of successful projects.

European Union accession: implications for environmental policy

Most CEE countries have declared their desire to become Member States of the European Union. The accession of CEE countries to the EU will, no doubt, have a tremendous impact on the CEE countries in all spheres and also on the European Union itself. This process will be dynamic, as some fundamental elements of the European Union will have to be adapted to its recent and future enlargement, but also as the transition process in the CEE countries progresses over the upcoming years.

It is also clear that this process will present a formidable challenge for environmental policy. The main motivations of CEE countries to become EU Member States are not environmental in nature. However, the accession process will have direct effects on environmental policy (e.g. alignment of environmental legislation and institutions). It will have huge, indirect impacts as well through the changes occurring in various sectors such as energy, transport, agriculture, and industry.

Today, there is little evidence that environmental policy will be a true priority issue for the CEE governments or that sustainable development is the long-term aim of the process of EU association. Concern and insecurity over the costs associated with the alignment to EU environmental regulations seem to be more pressing issues. The EU could be more of a driving force for increased respect of environmental policy if, for instance, the EU's fifth Environmental Action Programme 'Towards Sustainable Development' were to play a major role in the accession process. Nevertheless, the accession to the EU does represent a major opportunity for CEE governments to more deeply integrate environmental protection and economic development *in the process of* formulating new policies and establishing new structures.

Europe Agreements and Structured Relationship

Hungary and Poland signed association agreements with the EU in 1991 (these agreements came into force in 1994), while the Czech Republic, Slovakia, Romania and Bulgaria signed similar association agreements in 1993 (in force since February 1995). The three Baltic states, Estonia, Latvia and Lithuania, signed association agreements in 1995–96. Finally, after resolving a bilateral dispute with Italy, a Europe Agreement was signed with Slovenia in 1996. The second main instrument in the preparation process for EU accession is what is called the 'structured relationship' between the associated countries and the institutions of the EU, which complements the bilateral association agreements with a multilateral framework for strengthened dialogue and consultation.

The Europe Agreements provide the framework within which the countries can prepare for eventual accession to the European Union, and they include clauses outlining the actions that must be initiated in the environment area. Europe Agreement Committees meet regularly to discuss progress in these issues of approximation and harmonization. The Europe Agreements cite the PHARE Programme as a major instrument in helping to achieve the environmental preconditions for membership. (The PHARE Programme and its environmental component are discussed in more detail below.) For example, Poland's Europe Agreement states that the Polish Government and the EU, operating through PHARE, will co-operate in the following areas (European Commission, 1995a):

- Harmonization of environmental legislation and standards with those applicable in the EU, in close co-operation with the European Commission's Directorate General for Environment, Nuclear Safety and Civil Protection (DG XI).
- Pollution monitoring.
- Multi-country and cross-border pollution issues.
- Energy efficiency.
- Handling, storage and disposal of hazardous products and waste.
- Water quality, especially where bodies of water straddle borders or flow into the open sea.
- Agricultural practices and land management.
- Development of appropriate economic and financial instruments.

The White Paper and Mechanisms for the Approximation of the EU's Environmental Legislation

The Essen European Council of December 1994 adopted a broad pre-accession strategy for the associated countries of CEE. It identified preparation for integration into the EU's internal market as 'the key element in the strategy to narrow the gap' and invited the Commission to prepare a White Paper in co-operation with the associated CEE countries. This White Paper, 'Preparation of the Associated Countries of Central and Eastern Europe for Integration into the Internal Market of the Union', was approved by the Commission in spring 1995 (see European Union, 1995). The White Paper provides the conditions to be met that will allow the EU's single market to function properly to the benefit of all EU Members after enlargement.

The step-by-step implementation of the White Paper is likely to result in an alignment of competition and trade policies, and industry, agriculture, transport and energy policies. This will certainly influence and define the challenges for environmental policy during the accession process and after it, if this process eventually results in membership. The particular importance of the White Paper for environmental policy development in the associated countries of CEE must also be seen in this light.

It is stressed in the White Paper that the approximation can only be carried out by the associated countries themselves. Each of those countries will need to draw up its own programme of priorities and determine its own timetable according to its economic, political and social realities and the work it has achieved so far. It is also emphasized that the White Paper is one (but crucial) element in the pre-accession strategy and is not part of the negotiations for accession. Accession negotiations will cover the whole *acquis communautaire*, i.e. the whole field of European Union legislation and policy.

The White Paper also states that the EU's technical assistance which should catalyse the necessary alignment with the conditions set out in the White Paper will be carried out mainly through the PHARE Programme. For this purpose, the PHARE Programme will be enhanced and adapted. It is emphasized in the White Paper that it is the responsibility of each associated country to co-ordinate its requests for assistance and to provide information about progress made in implementing the White Paper, so as to ensure that assistance from the Union, Member States and other bodies is consistent and mutually reinforcing.

The White Paper contains (in a sector-by-sector approach) a listing and description of the legislation which is essential for the functioning of the internal market and also describes the administrative and organizational structures which are required in each sector to ensure that the legislation can be effectively implemented and enforced. The realization of this task in most CEE countries will not be easy and might be more time consuming than many wish. Throughout the region the legislative burden of parliaments is huge, the subject matter is sometimes unfamiliar and amendments to texts are numerous. In addition, recent experience in many CEE countries shows that governments and political priorities may change relatively quickly and profoundly, making a consistent transition process very difficult (European Union, 1995).

With respect to environmental policy, the White Paper mentions that 'environmental policy and the internal market are mutually supportive. The EU Treaty aims at sustainable growth and high levels of environmental protection and that environmental requirements be integrated into the definition and implementation of other policies. An integrated approach to allow a more sustainable path of social and economic development is not only vital for the environment itself, but also for the long-term success of the internal market.' (European Union, 1995). The White Paper also explains that the implementation of high common standards of environment protection is among the Union's objectives, as 'competition could be distorted if undertakings in one part of the Community had to bear much heavier costs [related to measures of environmental, social and consumer protection] than in another and there would be a risk of economic activity migrating to locations where costs were lower' (European Union, 1995).

While the importance of measures in the field of the environment is emphasized at various places in the White Paper, approximately 80% of the some 200 legislative pieces of the EU's environmental *acquis* is not covered in the White Paper, as for instance, legislation concerning quality standards for air and water and nature protection. Waste strategy is only covered to a limited extent. The reason for this is explained as follows: 'The White Paper includes legislation that directly affects the free movements of goods and services, leaving out legislation which relates to pollution from stationary sources and to processes rather than products and which therefore relates only indirectly to the internal market. [. . .] The present exercise concerning the internal market will therefore need to be complemented by a more comprehensive approach in the environment field, which is an important objective in its own right.' (European Union, 1995).

What this 'more comprehensive approach' entails should be elaborated in 1996. The European Commission prepared a concept for the preparation of the associated CEE countries for the approximation of the EU's environmental legislation in 1996 in co-operation with the associated CEE countries (European Commission, 1996). In light of the large and complex environmental *acquis* of the EU, the identification and elaboration of priorities of each associated country is proposed, in order to ensure successful and gradual approximation of legislation. It is for the associated CEE countries to determine their respective priorities and timetable, according to the needs and problems they face and the cost–benefit factors of action or inaction. A list of main criteria is proposed for prioritization (modified after: European Commission, 1996):

–   Legislation that can relieve imminent dangers to human health (poor air quality, poor quality of drinking water and groundwater, inadequate management of waste and hazardous substances).
–   Legislation to protect species or habitats of European or international importance (prevention of negative effects from land privatization, the expansion of agricultural production, and rapid economic growth).
–   Legislation that controls significant transboundary pollution.
–   'Framework' legislation, i.e. legislation items which provide the foundations for subsequent measures (regulation on EIA and free access to information on the environment are especially mentioned here).
–   Legislative and policy measures to build political and public support for environmental protection.
–   Legislation with a significant impact on the EU's internal market but not covered in the White Paper (e.g. regulation of industrial emissions for air and water, and the disposal of waste).
–   Implementation of existing international obligations.

It is stressed in the concept that the desired result of the approximation must not simply be the adoption of compatible legislation but also the realization of necessary steps on all levels to ensure the effective implementation and enforcement of the legislation. While it is clear that the key measures must be taken by the associated CEE countries, the EU provides support through the PHARE Programme, particularly through a special PHARE facility (PHARE Support Facility for Approximation of Environmental Legislation) which should provide 'the key assistance mechanism of cost/benefit analysis and specialized technical co-operation and advice'. It must be underlined, however, that the adoption of EU compatible legislation and the introduction of effective implementation and enforcement systems is a huge and difficult task and quick successes cannot be expected.

The PHARE Programme

The PHARE Programme is a European Union initiative. It supports the new democracies of Central and Eastern Europe and aims to assist these countries to join the recent European development and build closer political and economic ties with the European Union. PHARE pursues its goals by providing grant finance to support the process of economic transformation to market economies and the process of political transformation to democracies. Assistance through PHARE is also crucial to help prepare the countries with Europe Agreements for integration into the European Union, particularly also with respect to the requirements outlined in the White Paper on the preparation of the associated CEE countries for integration into the internal market of the Union.

The PHARE Programme became operational in 1990. As the name PHARE indicates (Poland and Hungary Assistance for the Reconstruction of the Economy), initial support was limited to Poland and Hungary. However, in late 1990, the group of partner countries was enlarged with Bulgaria, the former Czechoslovakia, the former GDR and the former Yugoslavia. In 1991, Albania and Romania were included as well, while assistance to the former Yugoslavia and the former GDR ceased. Finally, in 1992, Estonia, Latvia, Lithuania and Slovenia also became beneficiary countries.

In the first five years of operation, 1990–94, PHARE made available a total of 4248 million ECUs to the partner countries in Central and Eastern Europe. Table 1.6 shows the distribution by sector of PHARE funding in the period 1990–94. PHARE provides consulting services and policy advice from a wide range of public and private organizations to its partner countries. It provides training, and stimulates investments through studies, capital grants, guarantee schemes and credit lines. PHARE also invests directly in infrastructure which helps the progress of the restructuring process. It is the aim of PHARE that its programmes be defined and implemented in close co-operation with partner countries according to their individual needs, in order to ensure that the PHARE programmes are relevant to each government's own reform policies and priorities.

PHARE Environment Programmes

The initial environment programmes of PHARE addressed immediate problems and emergency environmental support projects. The 1990 projects in Poland, Hungary and the former Czechoslovakia concentrated on technical

Table 1.6   PHARE funding commitments by sector, 1990–94

Sector	percentage
Private sector development and enterprise support	21.9
Education and health	16.4
Infrastructure (energy, transport and telecommunication)	13.8
Environment and nuclear safety	9.4
Agricultural restructuring	9.4
Humanitarian and food aid	8.7
Public institution and administrative reform	4.9
Social development and employment	3.1
Others	12.3
Total PHARE funding commitments 1990–94 = 4248.4 million ECUs	100

Source: European Commission, 1995b.

feasibility studies and pilot projects in areas with an urgent need for improvement, provision of equipment for pollution monitoring, wastewater treatment and pollution reduction in power plants, and advice for institutional development, strategy development, and the like.

With the growing and more detailed knowledge about the true dimensions of environmental problems in CEE countries, and the evolving framework of environmental policies, institutions and legislation in these countries and on the international level (especially the Environment for Europe process), the emphasis of the PHARE programmes changed from 1991 to 1994. Also, programme implementation procedures started to be decentralized from Brussels to the partner countries. From 1991 to 1994, the following were the main areas of support in the PHARE environment programmes:

- Sectoral and feasibility studies addressing specific environmental issues, e.g. the management of hazardous waste, the treatment of wastewater, etc.
- Environmental policy and management including training, institutional strengthening, environmental strategy and action plan design, the development of enforcement systems, legislation and harmonization with EU and international standards.
- Equipment supply, action plans and pilot projects based on sectoral and feasibility studies.
- Ongoing pollution monitoring projects providing data which are critical for effective policy formulation and for enforcing compliance with permit systems.

Table 1.7 shows the amount of funding provided by PHARE environment programmes in the period 1990–94. While the share of environment pro-grammes in the Visegrad countries and Bulgaria was between 9 and 15% of total PHARE funding, this percentage was low in most other countries.

Table 1.7   PHARE environment and nuclear safety programmes: funding commitments 1990–94 (million ECU)

PHARE[a]	1990	1991	1992	1993	1994	Total	% from total
Czech Republic	0	0	0	0	0	0	0 %
CSFR	30.0	5.0	0	–	–	35.0	15.0%
Hungary	27.0	10.0	10.0	0	15.5	62.5	12.7%
Poland	22.0	35.0	18.0	0	12.0	87.0	8.6%
Slovak Republic	0	0	0	0	0	0	0 %
Bulgaria	3.5	26.5	8.3	10.8	5.0	54.1	13.7%
Romania	0	0	5.0	0	0	5.0	0.9%
Slovenia	0	0	0	0	0	0	0 %
Lithuania	0	0	0	0	1.0	1.0	1.2%
Latvia	0	0	0	0	5.5	5.5	8.8%
Estonia	0	0	0	0	2.5	2.5	5.6%
Albania	0	0	0	3.3	0	3.3	1.4%
ex-GDR	20.0	0	0	0	0	20.0	57.1%
Regional	0	23.5	46.0	20.0	36.0	125.5	17.4%
Total	102.5	100.0	87.3	34.1	77.5	401.4	9.4%

Source: European Commission, 1995b.

Note: [a]Percentage of environment and nuclear safety programmes as percentage of the total PHARE assistance to a country in 1990–94.

Apart from the various positive achievements in the implementation of the PHARE Environment Programme, it should be noted that this programme has also been criticized. In a resolution submitted to the European Parliament in early 1994 (European Parliament, 1994) which evaluated the PHARE Environment Programme in the Visegrad countries, a number of proposals for improvement were made. With respect to the management of the programme, the internal procedures of the programme were judged inadequate to ensure timely implementation of projects. Insufficient information and a lack of transparency on the evaluation of the initial period of the programme were also criticized. Furthermore, it was stressed that co-ordination within the overall PHARE Programme should be improved, i.e. that the environmental impact of important projects in other sectors (agriculture, energy, transport, etc.) should be assessed.

It was also noted in the resolution that consultants from the EU countries involved in implementing the projects often lack sufficient knowledge on the situation and practices in the CEE countries. Consequently it was proposed that local experts should be involved much more. In addition, it was suggested that exchange programmes of environmental experts should not be limited to officials, but be expanded to include technical experts and NGOs. It should be noted at this point that, in fact, the problems mentioned above are common to practically all assistance programmes, multilateral and bilateral (see section below on 'Problems with foreign and international assistance').

PHARE's Environment Strategy to the Year 2000

In the progress and strategy paper 'Environment to the year 2000', issued in February 1995, the strategic aims of the PHARE environment programmes for the next years are outlined. According to this document, PHARE and its CEE partners identified the focus areas for support as shown in Box 1.2.

The strategy paper also mentions a number of new procedures and instruments to further increase the effectiveness of the programme. The introduction of a fully decentralized implementation system was already mentioned. In addition, there are plans to introduce a 'policy advice' service which involves senior EU government and Commission officials advising their CEE counterparts on issues facilitating eventual accession to the EU. On future PHARE capital investment projects, an Environmental Impact Assessment will be required. Another important step is that the implementation of environmental investment projects will increasingly be channelled through national environmental funds of the CEE countries.

The strategic goals of the PHARE Environment Programme as outlined above are indeed ambitious and present a formidable challenge, but also underline the comprehensive importance of the Environment for Europe process for European policy development and convergence. Along with progress in association to the European Union, it is obvious that the PHARE Environment Programme will play an increasingly important strategic role with respect to CEE environmental policy development. However, it remains to be seen how effectively the programme can be implemented with limited funds and the available capacities.

1.3 FINANCING ENVIRONMENTAL PROTECTION IN CEE COUNTRIES

As information about the poor, and sometimes devastated, environmental conditions in CEE countries became more accessible and comprehensive following the collapse of those countries' centrally planned governments, there has been a widely perceived need for increased, and more effective, environmental investments in the region. This need has been analysed and reflected in a large number of studies, documents and gatherings on the local, national and international stage. As mentioned earlier in this chapter, the 'Environmental Action Programme for Central and Eastern Europe' (World Bank, 1994) identifies environmental investments as one of the three key areas (along with policy reform and institutional strengthening) for reform and improvement in environmental protection efforts in CEE countries. The recent Ministerial conference in Sofia (October 1995) re-affirmed, and perhaps

Box 1.2   PHARE environment strategy to the year 2000

Policy reform and harmonization
- Development of implementable national environmental action plans.
- Development of environmental policies which represent the goals of the European Union's fifth Environmental Action Programme 'Towards Sustainable Development'.
- Refinement and harmonization of environmental legislation with that of the EU.
- Clarification of environmental liability in the process of privatization.

Capital investments and environmental financing strategies
- Cost-efficient capital investments in the main areas of water pollution (especially waste water treatment), air emission reductions from industrial enterprises, and preservation of biodiversity.
- Introduction of a new cross-border co-operation programme.[3]
- Ongoing support and co-operation of/with environmental funds and eco-banks in CEE.
- Ongoing support of facilitating capital investments from international financial intermediaries.
- Further support to widen the environmental revenue base in line with the 'polluter pays' principle and to improve revenue collection systems.
- Support for new financing mechanisms in the environment (revolving credit facilities, etc.).

Institutional development and public awareness
- Encouraging inter-ministerial communication and co-operation to improve the integration of environmental concerns with the economic restructuring and reform process (especially in the areas of energy, agriculture, industry and finance).
- Development of stronger links with municipal authorities who have responsibility for the provision of environmental services like drinking water supply and sewerage facilities.
- Continuation of training for senior managers and policy-makers in the private and public sectors in planning, management, financing, forecasting and policy-making.
- Improvement of the effectiveness of enforcement agencies through training and advice and the installation of monitoring and analysis systems.
- Raising the environmental awareness of the public but also among enterprises and business to strengthen environmental policy-making.

Source: modified after European Commission, 1995a.

even amplified the importance of enhanced financing for environmentally beneficial investments in the region (Sofia Declaration, 1995).

This emphasis on the issue of financing is at least partly the result of dissatisfaction among CEE governments with the efforts of Western donors and international financing institutions, who have sometimes been perceived as slow in fulfilling their commitments made under the EAP. There has been some sense in CEE countries that, in the past, too many resources were consumed by the preparation of feasibility and other studies, and advice on policy reform and institutional strengthening, but not enough money invested in the completion of projects yielding direct and measurable environmental improvements. There were clearly some unrealistic expectations within the CEE countries as well as misunderstandings on the part of both donors and recipients as to what were the commitments of each. Besides misunderstandings and unrealistic expectations, however, simple reality has required a new focus for environmental financing initiatives in the region. The reality is that the CEE countries themselves are bearing, and will continue to bear, the large majority of costs for environmental protection measures in their nations (COWIconsult, 1995; Laurson et al., 1995). This will come as no surprise to many people, both within and outside of the CEE region, and it is only natural that any given country or region takes principal responsibility for solving its own problems. The clarity of the situation has prompted the development of strategies for environmental financing in the region which emphasize maximizing the effective and efficient use of *domestically generated* resources, rather than the overall *supply* of resources (whether domestic or foreign). This is not to say that more resources for environmental investment are not being sought, they are, nor that foreign sources of financing are unimportant, they are and will remain so, especially for their leveraging and catalytic effect.

The various strategies and efforts being made to improve the financing of environmental investments in CEE countries are confronted with numerous impediments. Several of the more significant of these are analysed in the following section of this chapter. Subsequently, information about CEE country domestic sources as well as Western and international sources of environmental finance are presented. The section concludes with a brief discussion of emerging trends and innovative mechanisms for financing environmental protection in the region.

Impediments to financing environmental protection in CEE countries

The problems involved with financing environmental protection measures in CEE countries do not result solely from those countries' lack of financial resources, or from the chaotic times they are passing through during their transitions to democratic, market economies (though some of the most important problems certainly are closely related to these phenomena). There is a wider range of impediments to financing environmental protection in the region. An excellent discussion of these impediments can be found in a paper prepared by the Harvard Institute for International Development's Central and Eastern Europe Environmental Economics and Policy Project (HIID, 1995). The paper rightly explains a few caveats concerning the issue before discussing the impediments themselves. As the authors of that paper noted, when considering obstacles to greater environmental investments in the region it is important to remember the huge, unsatisfied need and demand for investments in many sectors within the CEE countries, and that the environ-

---

Box 1.3   Fundamental impediments and estimating environmental expenditures

The first two impediments to environmental investments discussed above – low public and political support for environmental protection and the poor financial condition of governments, enterprises and citizens – are perhaps the most fundamental of all the obstacles to greater environmental expenditures in the region, both for the short and long term. The financial hardships being experienced throughout the region during the transformation process render any dramatic increases in environmental expenditures at the national or regional level very unlikely for the near future. (Hence the growing emphasis on efficient use of available resources.) As for the longer term, assuming that the CEE countries progress through the transformation process without any traumatic setbacks (i.e. war or political upheaval), public and political support will play a major role in ensuring that the levels of their expenditures on environmental protection rise to approximately the levels of such expenditures in Western countries, where current public and political support for environmental protection is generally believed to be higher.

Though many of the CEE countries have made considerable strides in environmental protection in recent years, to this point most of them have not been able to sustain environmental investment expenditure at levels comparable to those typical of OECD countries, generally 1–2% of GDP (Laurson et al., 1995). The calculation of environmental expenditure as a percentage of GDP is no straightforward affair, however, and use of this indicator to assess a country's 'environmental commitment' must be qualified. First of all, it is sometimes difficult to define precisely what an 'environmental' expenditure is. Some types of expenditure included in such calculations by some countries may have only marginal environmental benefits, while some may even be more environmentally-unfriendly than beneficial. Different CEE countries use different definitions of 'environmental expenditures', thus some report misleadingly high figures and others conservatively low figures. The upshot of it all is that environmental expenditure as a percentage of GDP is an elusive figure at best and very difficult to use reliably in making comparisons among countries in the region and between CEE and Western countries.

---

ment sector represents but one sectoral demand on the scarce available resources. Additionally, it is difficult to assess accurately just what the current levels of environmental expenditure being made in CEE countries are and, hence, to know how they compare with other countries in the region or Western countries (see Box 1.3). With these issues in mind, a number of impediments (though certainly not all) to financing environmental protection in CEE countries are analysed below.

Low Support for Environmental Protection

As was described in the first section of this chapter, support for environmental protection activities, and the expenditure they necessitate, is generally low in CEE countries. Public pressure for strengthened environmental protection is weak and this translates into low interest and support among politicians and other influential decision-makers. Environmental issues, once rallying points for anti-Communist protests and critiques, are now virtually off the political and social agendas in CEE countries. In the wake of perceived victory over the old, common political foe, and with new, intense and previously unknown hardships being experienced as a result of social and economic transformation, most CEE citizens are now far more concerned with feeding their families, paying the rent and keeping their jobs than with improving environmental conditions. Of course, there are pro-environment lobbies in each of the CEE countries, and they are gradually becoming more professional and effective, especially as public participation in decision-making becomes more of a reality than a catchy phrase, but this lobby remains weak compared to other social and commercial interest groups. Until this situation changes significantly it seems unlikely that the public opinion and policy and institutional framework necessary to produce substantially increased environmental investments will occur.

Poor Financial State of Governments, Enterprises and Citizens

With the collapse of the Soviet trading block, COMECON, the onset of 'hard' budgets, the opening of borders for competitive trade and subsequent economic recession, most governments in the region (national and local), and many large and medium-sized enterprises, especially in the heavy (and most-polluting) industries, are suffering serious financial constraints which affect investment expenditures of all kinds, including environmental. Half or more of the largest polluting companies in CEE countries may be financially insolvent (HIID, 1995). Environmental expenditure may be even more seriously con-strained as such investments generally have negative net cash flows or are only marginally profitable and investors (private and public sector) usually prefer to use their money for more productive activities. Citizens of CEE countries are also feeling the economic pinch and often consider environmental benefits as a luxury or of secondary importance.

Over-centralization of Government Decision-making and Constraints on Local Governments

Even with ongoing decentralization in many CEE countries, decision-making often remains largely in the hands of national government officials located in the state capital. Environmental threats and impacts, however, are experienced mainly at the local level and in highly centralized decision-making systems local authorities and citizens tend to have little influence over the allocation of state government resources. Thus, decisions made at the central level by the Ministry of Finance may not accurately reflect needs and demands at the local level (HIID, 1995). Additionally, while decentralization has transferred some new responsibilities and authorities to local governments, these governments often do not have the legal rights to undertake certain forms of financing activities. National public finance laws make it difficult for municipalities in many CEE countries to raise the funds they need (through bonds, for example) to finance the environmental services that they are now responsible for providing.

Unstable Macro-economic Conditions and an Uncertain Regulatory Environment

In some CEE countries macro-economic conditions are still unstable. Under such conditions investors of all types tend to be very wary given the high risks involved. Moreover, serious economic fluctuations or high inflation can easily undermine investment incentives which might have been created by positive reforms in environmental policy or improvements in institutional arrangements. Even in those CEE countries where macro-economic conditions have stabilized, there remains considerable uncertainty with regards to future environmental standards. The environmental regulatory systems are evolving in all of the CEE countries, with old laws being reformed or sometimes entirely replaced. The pace of evolution varies considerably from country to country and even within countries according to specific law or environmental sector. Generally speaking, the market is developing more quickly than the regulatory regime, resulting in pressures on governments and enterprises to act (i.e. make investments) often without sufficient knowledge as to what standards they will be required to enforce or comply with in the future. The situation is further complicated by the fact that many of the existing environmental standards, especially those remaining from the Communist era, are unrealistically strict and impossible to implement. While new regulations may be coming down the pipe to replace the old, how do enterprises know what actions to take, investments to make, in order to be 'in compliance'? Given that many of the CEE countries have expressed their interest in joining the EU, EU standards offer guidance as to what the future standards might look like. Nevertheless, the harmonization process will not be accomplished over night and the uncertainty will continue in many of the countries for some time to come.

Weak Economic and Regulatory Incentives

A number of economic and regulatory incentives which should encourage environmental expenditure in CEE countries are simply not as effective as they should be. The first is actually a 'disincentive': the prices of energy sources and other primary material inputs are still subsidized (often heavily) in CEE countries, thus encouraging inefficient and wasteful use of natural resources and the forgoing of investments which would simultaneously reduce resource consumption and pollution emissions. Energy and raw materials prices are, for the most part, gradually being rationalized; however, the process is incremental and strong political lobbies exist in certain industrial sectors (i.e. coal) which oppose the process. Additionally, economic instruments such as pollution charges and fines, which should serve as incentives for enterprises to change polluting behaviour, do not fulfil this function because the levels are too low in most CEE countries. While such charges and fines do generate revenues which are often used for environmental protection purposes (i.e. through environmental funds), the charge and fine levels are generally so minimal that it is less expensive for the polluting enterprises simply to continue paying them rather than to make investments which would eliminate or reduce emissions. Moreover, some enterprises do not pay charges or fines at all due to financial insolvency, privileged status *vis-à-vis* regulators, simple non-compliance, etc. Enforcement of environmental laws and regulations remains a major weak point in the environmental protection systems of these countries. As mentioned, some regulations are unrealistically strict and impossible to meet, while the enforcement agencies, typically the environmental inspectorates, often lack the resources and political support necessary to do their jobs properly.

Insufficient Capacity to Prepare Investment Projects and Difficulties with Commercial Capital

It is widely believed that the capacity existing in the region to identify, prepare and package environmental investments for financing is underdeveloped (UNECE, 1995a; HIID, 1995) and that viable projects go unimplemented as a result. People with strong technical and engineering skills are not in short supply; however, the preparation of environmental investments and the securing of financing for them requires additional skills which are not yet commonly found in the CEE countries, such as cost-effectiveness analysis, business planning and financial/credit analysis. A variety of deficiencies within the commercial capital market also present obstacles to financing environmental investments in the region. The banking sectors of many CEE countries are still underdeveloped and undercapitalized, unable or unwilling to extend medium- to long-term loans at affordable rates (if at all), and inexperienced with environmental investments and unwilling to assume the perceived risks associated with such investments (Laurson et al., 1995). The basic result is that commercial capital is often not available for environmental investments (which typically require longer payback periods than other types of investments and have lower rates of return), or is prohibitively expensive for potential borrowers.

Domestic sources of environmental financing in CEE countries

As stated in the introduction to this section, the CEE countries themselves are providing the lion's share of financing spent for environmental protection activities in their countries (see Table 1.8). The main sources of this financing

Table 1.8   Total environmental expenditures in selected CEE countries for 1994 with shares contributed by domestic and foreign sources, and per capita estimates (US$)

Country	Domestic financing	International loans/grants	Total expenditure[a]	Expenditure per capita[4]
Bulgaria	96%[1]	4%[1]	103 million[2]	12
Estonia[1]	65%	35%[b]	68 million	45
Poland[3]	96%	4%	936 million	24
Slovakia[1]	98%	2%	212 million	40

Sources: [1]COWIconsult (1995); [2]Popova (1996), does not include foreign sources; [3]Estimates of the Polish Ministry of Environment (1995), does not include expenditure on water supply systems, forestry or operation and maintenance of investments); [4]Author's calculations.

Note: [a]The reader should note that 'environmental expenditures' is defined differently in different CEE countries, with some countries having more inclusive definitions than others, thus rendering comparison between countries difficult and possibly misleading. [b]Estonia has received sizeable grants from neighbouring Finland.

Table 1.9   Profile of domestic environmental financing sources in selected CEE countries for 1994 (as a percentage of total domestic environmental expenditure)

Country	State budget	Regional and municipal budgets	Extra-budgetary funds[a]	Enterprises own resources
Bulgaria[1]	21	8	5	66
Estonia[1]	21	3	8	68
Poland[2]	5	20	43	32
Slovakia[1]	51	16	16	17

Source: [1]Calculations based on a report prepared by COWIconsult (1995); [2]Author's calculations based on estimates of the Polish Ministry of Environment (1995).

Notes: [a]Includes national and sub-national environmental funds.

tend to be the central state budgets, the private resources of enterprises and extra-budgetary environmental funds (see Table 1.9). The significance of the roles being played by these different institutions, however, is changing. As a result of severe fiscal constraints and decentralization, the role of the state budget in environmental financing has decreased in some of the countries since 1990 (Francis, 1994; COWIconsult, 1995). This trend will almost certainly continue, though the state budget is likely to remain an important source of environmental financing in some CEE countries for much of the transition period. With a decline in the role of the state budget, the share of environmental expenditure contributed by local government and environmental funds (national and sub-national) has increased, and is increasing, in some countries. The role of polluting enterprises in environmental expenditure will also increase as the 'polluter pays' principle becomes more fully adopted and implemented in the region. (Admittedly, this may be a slow process, as numerous polluters continue to be granted exemptions and special privileges by the authorities (Laurson et al., 1995), thus allowing many polluters to evade their responsibilities for environmental protection, resulting in those responsibilities being shifted back on to the public.)

Extra-budgetary Environmental Funds

Extra-budgetary environmental funds are receiving growing attention as mechanisms for providing, leveraging and facilitating finance for environmental protection activities in CEE countries. These funds are quasi-independent or independent institutions, having been created by the initiative of Ministries of Environment for the purposes of providing additional, earmarked finance for the support of environmental protection activities (Francis, 1994). Such funds exist, or are in the process of being formed, in nearly every CEE country. (Exceptions include Romania and some of the Balkan states.) In almost all of the countries where they function, these funds are operated under the auspices of the national Ministry of Environment. (The major exception is Poland, which has a national and 49 provincial funds, existing as legally independent institutions created by Parliamentary Act.) The funds typically receive revenues from pollution charges and fines, environmental taxes, product charges and other fees on the use of natural resources and the environment. The funds then use this money to support environmentally beneficial activities, such as investments in pollution control and prevention technology, environmental education, and the establishment of environmental monitoring systems. The financial support provided by the funds is disbursed in various forms, most commonly as grants and soft loans. Table 1.10 summarizes the key characteristics of selected national environmental funds in the region.

As mentioned, and reflected to some extent in Table 1.11, the importance of the funds in environmental expenditure is very considerable in some, and growing in most, CEE countries. While Table 1.11 covers selected national funds, there are also other environmental funds in the region which deserve attention. Lithuania, Bulgaria and Poland have environmental funds at the municipality/community level, though these funds tend to be rather small and their expenditure may be only marginally environmental in nature. More significant, however, are the 49 provincial funds located in Poland. The largest of these funds have revenues in the tens of millions of dollars, making them comparable in size to, or even larger than, some of the national environmental funds in other CEE countries. (Poland's national and provincial environmental funds account for over 40% of the country's environmental expenditure.) A particularly unique environmental fund in the region is Poland's 'EcoFund'. This fund was established as an independent foundation by Poland's Ministry of Finance to administer revenues resulting from the forgiveness of national debt by a few of Poland's creditor nations. To this

Table 1.10 Key characteristics of selected national environmental funds in CEE countries

Key characteristics[1]	Bulgaria	Czech Republic	Hungary	Poland	Slovak Republic
Operational since (in current form)	1993	1991	1993	1989	1991
Institutional status	Part of MoE	Part of MoE	Part of MoE	Independent	Part of MoE
Final decision-making authority	Minister of the Environment	Minister of the Environment	Minister of the Environment	Supervisory Board (for projects > 200 000 US$); Fund Directors (otherwise)	Minister of the Environment
Advisory role played by:	Board of Directors and Supreme Board of Environmental Experts	The National Environmental Fund Board	Interministerial Committee	Supervisory Board	Fund Council
Public participation opportunities in decision-making	Included in the work of the Supreme Board; public hearings twice a year	During environmental impact assessments	NGOs are consulted during the formulation of the Annual Programme for Support	Through the membership in the Supervisory Board	Through the Fund Council
Basis of long-term spending strategies	National Environmental Policy	National Environmental Policy	Conceptual settings of MoE	National Environmental Policy	National Environmental Policy
Major revenue sources (1994)[2]	– pollution fines – import tax on used cars – contribution from privatization	– air charges – water charges – land charges – waste charges	– fuel tax – mining fees – pollution fines – investment interest – PHARE support	– air emission charges – wastewater charges – water use charges – waste charges	– air emission charges – state budget – wastewater charges
Primary disbursement mechanisms	– grants – interest-free loans	– grants – soft loans	– grants – interest-free loans – loan guarantees	– soft loans – grants – interest subsidies	– grants

Sources: [1]Francis (1994); [2]Bulgaria, Slovakia, and Poland – COWIconsult (1995); Czech Republic – Stepanek (1995); Hungary – Lehoczki (Chapter 4 of this volume).

Table 1.11   Environmental expenditures of selected national environmental funds in CEE countries, 1993–95 (millions of US$)

Year	Bulgaria	Czech Republic	Estonia	Hungary	Poland	Slovak Republic
1993	2.3[1]	107.0[1]	0.5[2]	27.7[1]	198.5[1]	34.7[1]
1994	2.5[3]	163.0[4]	1.7[2]	25.9[5]	224.5[6]	31.0[2]
1995	6.3[7]	–	4.4[8]	43.8[5]	437.3[9]	33.9[8]

Sources: [1]Francis (1994); [2]COWIconsult (1995); [3]Chabrzyk (1995); [4]Stepanek (1995); [5]Hungarian State Budget (1994, 1995); [6]Kruszewski (1995); [7]Popova (1996); [8]COWIconsult (1995) with June 1995 exchange rate; [9]Polish National Fund (1996) with 1995 average annual exchange rate.

point, the USA and Switzerland have agreed to allocate 10% of Poland's debt for environmental protection (US$370 million and US$53 million, respectively), while France has agreed to allocate 1% of Polish debt to the fund (US$50 million) (Nowicki, 1995). (See Chapter 6 in this volume for more discussion about the EcoFund.) Developments are also well advanced to create a similar 'debt-for-environment swap' fund in Bulgaria (discussed further in Chapter 2).

The expanding significance of the funds has been accompanied by an increased interest in their role in public finance, as well as scrutiny of their capacity to effectively and efficiently manage public moneys for the public good. Numerous meetings and conferences have been convened at the national and international level to discuss the role and activities of the funds. One of the most important of these gatherings was the Conference on Environmental Funds in the Transition to a Market Economy, organized by the OECD and held in St Petersburg, October 1994. This conference resulted in the St Petersburg Guidelines on Environmental Funds in the Transition to a Market Economy, which endorsed environmental funds as useful mechanisms for supporting environmental investments in CEE countries during the transition period. The guidelines also stress, however, that certain risks are associated with such funds and that the funds must be carefully designed, managed and evaluated. The guidelines state:

> Without careful design and management, however, the potential advantages of Funds could become defects. From a *fiscal policy perspective*, 'earmarking' has potential dangers: allocating and disbursing revenues outside the government budget may create long-term economic inefficiencies. Well-designed procedures and incentives are needed to ensure that Environmental Funds target priority environmental problems and that Fund revenues are spent effectively. This requires effective project appraisal techniques as well as financial and accounting procedures. Funds should ensure transparency and they should be accountable to government, Parliament, and the public for their actions . . .

> (Lehoczki and Peszko, 1995) (Original italics)

A variety of efforts are now underway within the CEE countries themselves and donor communities to evaluate the activities of the funds in order to help them improve their effectiveness and efficiency. A number of partnerships have already been formed between funds and Western assistance institutions and such relationships are likely to become more common as the funds improve their capacities in line with the St Petersburg Guidelines.

Emerging Mechanisms for Domestic Environmental Finance

The economies and financing systems of CEE countries are evolving and so too are the opportunities for them to finance environmental protection activities. CEE countries which have progressed the furthest in economic transition are beginning to make use of financing mechanisms commonly used in Western countries for the support of environmental investments. First of all, it should be remembered that in the West the commercial banking sector plays a major role in financing environmental investments. While significant obstacles and reluctance still remain, the commercial banking sectors of some CEE countries (namely the Visegrad group) are becoming more active in the environmental sector. Early steps are also being taken in the Czech Republic, Hungary and Poland with the use of 'green' equity to generate financing for environmentally beneficial investments at the enterprise and municipal level. 'Green' equity investments involve investors who are willing to invest money in a project or enterprise without a specific claim for direct repayment, but rather in expectation that the project or enterprise will, over time, become profitable enough to reward the investor. Though some kinds of environmental projects or enterprises are profitable, the payback periods may be too long to attract commercial investors. Therefore, it is likely that 'green' equity schemes will require the involvement of non-commercial investors, such as environmental funds, bilateral donors or international development banks.

Foreign and international financing of environmental protection in CEE countries

Although the above discussion focuses on the dominant role of domestic financing sources for environmental protection in CEE countries, the role of foreign bilateral and international sources of environmental finance remains important for a number of reasons. First of all, the legacy of past environmental neglect and abuse in CEE countries under the Communist regimes requires massive expenditure for environmental clean-up, abatement and pollution prevention – far more than CEE countries are able to handle at this time. Foreign and bilateral sources are certainly not able to bridge such gaps in environmental financing, but they do provide additional financial resources not otherwise available within the countries. Second, foreign and international sources of finance can play a strategic role in supporting and augmenting environmental expenditure in CEE countries. Foreign assistance can focus resources on those highest priorities identified by CEE countries and, if needed, help the countries to identify those priorities in the first place. Foreign aid can also serve as a catalyst for mobilizing additional finance (domestic and foreign) for environmental investments and programmes. This 'leveraging' effect is very important as most environmental investment projects in the region, especially large projects, require multiple co-financing sources. Foreign bilateral and international finance can also be extremely valuable in supporting demonstration and pilot projects, whose benefits go beyond the immediate direct effects of the individual project, for example, promoting replicability and/or the transfer/development of clean technology. Certain skills, methodologies and knowledge, critical for improving environmental protection in CEE countries, are also greatly needed in the region and foreign assistance programmes are sometimes the only mechanisms able to finance and facilitate the development of such capacities.

Since 1991 (and even earlier) there have been a great many foreign bilateral and international environmental assistance programmes and projects in the CEE countries, representing a very considerable sum of money (see Table 1.12). Of the numerous mechanisms and institutions involved in providing and facilitating foreign environmental assistance in CEE countries several stand out as being especially significant, such as the EU PHARE Programme (discussed earlier in this chapter), the World Bank, the European Bank for Reconstruction and Development, the European Investment Bank, the Global Environment Facility and numerous bilateral programmes involving many Western countries (i.e. United States, United Kingdom, Germany, Denmark, The Netherlands, Sweden, Switzerland, Austria, Norway, France, Japan and Finland). The environmental assistance commitments of IFIs, as well as the commitments of OECD countries and the EU to CEE countries during years 1991–94, are presented in Tables 1.13 and 1.14.

Problems with Foreign and International Environmental Assistance

Many of these foreign assistance programmes and projects have been very successful, some only marginally so, and still others heavily criticized as failures or worse. A number of initiatives have been undertaken recently to assess the effectiveness of international financial support for environmental protection in CEE countries, and several major deficiencies of such support were summarized by Klaassen and Smith (1995) as follows (based substantially on the input of CEE experts):

- Foreign aid has often been focused on the preparation of studies, particularly feasibility studies, with inadequate emphasis on actual project implementation which would result in real environmental benefits.

Table 1.12 Commitments of foreign environmental assistance and finance to CEE countries from 1991–94[a], according to recipient (millions of ECUs, except per capita)

Recipient country	Technical co-operation			Total[b] (million ECUs)	Total per capita (ECUs)
	Policy development	Investment preparation	Investments		
Albania	3.4	1.5	67.1	72.2	21.2
Bulgaria	15.8	5.7	268.4	291.2	32.7
Croatia	0.0	0.2	0.0	0.2	0.0
Czech Republic	20.0	42.3	298.0	361.4	34.4
Estonia	8.5	4.5	122.4	136.0	85.0
FYR Macedonia[c]	0.1	0.1	0.0	0.2	0.0
Hungary	32.2	5.5	201.2	239.9	22.8
Latvia	11.3	2.8	33.3	47.8	17.7
Lithuania	8.6	6.0	49.0	64.4	16.9
Poland	51.7	31.6	932.9	1017.9	26.4
Romania	10.4	4.1	176.0	190.9	8.2
Slovak Republic	15.4	12.4	142.1	170.4	31.6
Slovenia	0.8	15.3	13.2	29.2	14.6
CEE region-wide	46.4	27.6	5.1	79.9	–
Total CEEC	224.5	159.7	2308.7	2701.5	23.0

Source: UNECE (1995a).

Note: [a]Based on responses from Austria, Denmark, Finland, France, Germany, Japan, The Netherlands, Norway, Sweden, Switzerland, United Kingdom, United States, European Commission, EBRD, EIB, GEF, NEFCO, Nordic Investment Bank. World Bank data are only partial. [b]Totals may be larger than sum of technical co-operation and investment assistance, as some donors did not classify commitments. [c]Former Yugoslav Republic of Macedonia.

- Moreover, the feasibility studies themselves often did not pay enough attention to financing issues and thus project design sometimes posed obstacles to financing.
- Foreign loan proposals have sometimes failed to take into account the need for state guarantees, which might not be available due to fiscal constraints or policies of the state government.
- The needs and priorities of the donors do not always match the needs and priorities of the recipients. Additionally, sometimes the recipients have difficulties in identifying their own needs and priorities and donors should be aware of this and be prepared to help.
- Foreign aid is often 'tied', requiring recipients to purchase donor country equipment which might not be the preferred choice of the recipient, the most cost-effective or even appropriate technology.
- Similarly, a considerable portion of foreign assistance often goes to foreign consultants whose comparative advantages over local consultants may be marginal or even non-existent due to their relatively high costs and occasional lack of knowledge of the in-country situation.

Table 1.13   Environmental commitments of IFIs[a] to CEE countries from 1991–94
(millions of ECUs)

	1991	1992	1993	1994
Investments	359.83	286.42	389.94	864.00
Technical co-operation	3.84	7.68	9.01	2.68
Total	363.67	294.10	398.95	866.68

Source: EAP Task Force Secretariat–OECD (1995e).

Note: [a]Refers to data from the European Investment Bank, EBRD, GEF, NEFCO, Nordic
Investment Bank and World Bank. (World Bank data are only partial.)

Table 1.14   Environmental commitments of OECD countries and the European
commission to CEE countries from 1991–94 (millions of ECUs)

	1991	1992	1993	1994
Investments	9.32	8.29	35.95	69.14
Technical co-operation	40.30	75.75	53.06	52.69
Total	49.62	80.04	89.01	121.83

Source: EAP Task Force Secretariat–OECD (1995e).

Note: [a]Based on responses from Austria, Denmark, France, Japan, The Netherlands, Norway,
Sweden, Switzerland, United Kingdom and the United States.

– Procedures for obtaining international assistance can be complicated, time
  consuming and expensive. Recipients sometimes are not well informed
  about such procedures and may lack the capacity or resources to deal
  with them.
– Assistance efforts are sometimes focused on projects which are either too
  small or too large to meet the needs of the recipient countries. International financial institutions generally have minimum project sizes which
  might preclude the involvement of numerous smaller projects, while
  bilateral programmes sometimes lack the resources necessary to address
  major investment needs.

Improving the Effectiveness of Foreign Environmental Assistance

The different problems listed above concerning the provision and use of
foreign environmental assistance to CEE countries may also serve as a partial
guide for how to improve the effectiveness of such assistance. More assistance
needs to be focused on actual investments, or the securing of financing for
investments, rather than on feasibility studies (acknowledging, however, that
such studies are important and useful *if* they lead to real environmental
benefits); the impediments to financing environmental improvements in CEE

countries need to be well understood and accounted for by donors and IFIs; donors and IFIs need to know and respect the CEE countries' priorities and, if necessary, help them to clarify those priorities; foreign aid should be untied to the maximum extent possible; local consultants should play major roles in the identification, design and implementation of assistance programmes; and donors and IFIs need to establish good communications with CEE countries and must be able to explain their expectations and requirements while taking into consideration the needs and capacities of their beneficiaries. This list of suggested improvements is very cursory, but it does include some of the major issues to be addressed.

As more attention is given to improving the effect of foreign and international environmental assistance for CEE countries, and as lessons are learned from experience to date, constructive efforts are being undertaken on the part of both donors and recipients. Figure 1.8 illustrates the same data presented above, but the line graphs clearly show the significant increase in commitments made by OECD countries, the European Commission and international financial institutions for environmental investments. (Of course, one must remember that there is always a difference between amounts committed and amounts actually disbursed.) To some extent this represents a natural progression from the investigations and feasibility studies which were necessary in the early years of substantial donor and IFI activity in the region as well as the general development of the countries which allowed greater investment opportunities. Special initiatives have, however, also been started with the explicit goal of bringing about the implementation of more environmental investments yielding greater environmental benefits.

The Project Preparation Committee

One of the more significant of these special initiatives has been the creation of the Project Preparation Committee. Known more commonly by its acronym PPC, this committee was established in 1993 within the framework of the Environment for Europe process. The PPC's main objective is to facilitate the preparation and implementation of environmental investments in CEE and NIS countries by improving co-ordination between donors, international financial institutions and the CEE/NIS countries (PPC, 1995). The PPC focuses its attention on environmental investments meeting the criteria of the Environmental Action Programme for Central and Eastern Europe, that is, actions which bring immediate or short-term relief to regions where human health or natural ecosystems are severely jeopardized by environmental hazards. At present the PPC includes representatives of 11 countries (Austria, Denmark, Finland, France, Germany, The Netherlands, Norway, Sweden, Switzerland, the United Kingdom and the United States of America); the EU's PHARE and TACIS Programmes (TACIS is the EU's counterpart to PHARE for the Newly Independent States); and five international financial

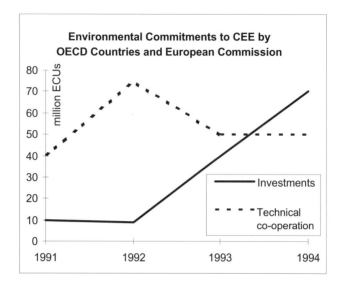

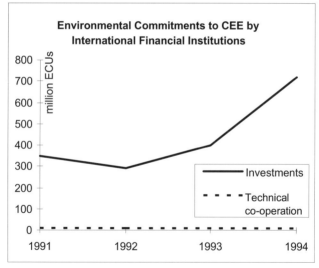

Figure 1.8   Environmental commitments to CEE countries during 1991–94 by OECD countries, the European Commission and IFIs (in millions of ECUs). *Source*: UNECE (1995a)

institutions (the European Bank for Reconstruction and Development, the European Investment Bank, the International Finance Corporation, the Nordic Environment Finance Corporation and the World Bank). Additionally, the Environmental Action Programme Task Force, the Global Environment Facility, the United Nations Development Programme and Japan have observer status on the committee.

The PPC is not a fund and does not have its own resources to finance environmental investments. Rather, it functions as a co-ordination network to enable IFIs and donor countries to identify and match environmental investment projects which have been given priority by CEE/NIS countries. Furthermore, the PPC assists CEE and NIS countries in developing domestic financial resources for environmental investments and provides a forum for sharing experiences between donor countries and IFIs. By mid-August 1995, the PPC had 'matched' 79 environmental investment projects in 21 CEE and NIS countries. From these projects, 26 have been approved by IFI Boards or funding agencies, or are already under implementation. The investment costs of the approved projects amount to 1.2 billion ECUs, of which assistance funding from donors amounts to about 80 million ECUs. In the future the PPC will continue its co-ordinating and facilitating role and has identified the following areas for special attention (PPC, 1995):

- Increasing the leveraging effect of scarce financial resources by contributing to the establishment of new multilateral financing mechanisms and the strengthening of existing ones.
- Improving the effectiveness of financial intermediaries and stimulating growth and capacity-building within the banking and financial sectors in CEE countries.
- Expanding and improving co-operation with environmental funds.
- Increasing and supporting the involvement of the private sector in environmental investments.
- Becoming involved in the identification and preparation of environmental investment projects at the earliest possible stage of the project cycle.
- Enhancing co-operation with the Environmental Action Programme Task Force and increasing the transparency of its own decision-making processes.

Emerging Mechanisms for Foreign and International Environmental Finance

The PPC and many other institutions have expressed strong interest in supporting the development of new, improved and innovative environmental finance mechanisms for CEE countries. All of these mechanisms will rely heavily on the integral involvement of the CEE countries. The emergence of the commercial banking sector and 'green' equity schemes in the financing of environmental investments was discussed earlier in the subsection on domestic finance. Other interesting schemes and concepts for enhancing environmental finance in the CEE region include: a loan guarantee scheme; a long-term environmental assistance facility; a consolidated programme supporting institutional strengthening for environmental funds; joint implementation; and new 'debt-for-environment' swap mechanisms. Each of these is briefly described below.

- *Loan guarantee scheme*   At present, some potential environmental invest-ments in CEE countries are unrealized because the guarantees required by lenders to ensure repayment of their loans are often difficult or impossible to obtain. This scheme would reduce the risk of lenders for environmental investments by channelling donor funds through appropriate in-country institutions to be used for loan guarantees (PHARE, 1995).
- *Long-term environmental assistance facility (LEAF)*   This proposed facility would basically function as a 'soft' green bank, providing very low-interest loans and grants for environmental projects. The facility would be associated with a European IFI, adhere to that IFI's standards for project identification, appraisal and selection and be financed by profits of the IFI and grants from its members (PHARE, 1995).
- *Programme for strengthening environmental funds*   The importance of the funds in financing environmental protection in CEE countries has been discussed in some detail. They are increasingly being looked to by Western and international partners as useful instruments for supporting and facilitating Western environmental assistance. Strategic, systematic and steady improvement in the operational efficiency of the funds would magnify the positive impacts of both domestic and foreign resources for environmental protection.
- *Joint implementation*   In the context of environmental investments, joint implementation refers to the situation where one country (donor) meets part of its obligations under an international environmental agreement by supporting investment action in the territory of another participating (CEE) country. The scheme is motivated by the fact that it tends to be less expensive for donor countries to achieve a certain level of pollution reduction by investing in actions in CEE countries than by investing at home (UNECE, 1995a). (It should be noted, however, that this approach has been strongly criticized by some, especially citizen environmental organizations, on the grounds that it allows the 'donor' to avoid its own responsibilities at home and thus merely shifts attention away from the real causes of the environmental problem the relevant agreement is trying to solve.)
- *Debt-for-environment swap*   Based on the very successful experience of the EcoFund in Poland, this mechanism is being promoted as an environ-mental financing mechanism which might be desirable and effective in some CEE countries. The approach does have its limitations, however, including significant fiscal implications for the debtor country.

All but one of these innovative approaches (the exception being the LEAF), plus the concept for a 'green' equity scheme were presented in the *Integrated Report on Environmental Financing*, prepared by the EAP Task Force Secretariat and PPC and submitted to the European Environmental Ministers Conference in Sofia, October 1995. The Ministers broadly endorsed these concepts (Sofia Declaration, 1995) and efforts are already underway to begin, or further, their implementation. As existing and new mechanisms for financing environmental protection in CEE countries evolve and mature, and as the countries themselves continue to transform and progress through their transitions, the responsibilities and costs for environmental protection will increasingly be borne, much more directly and fully, by the producers of the pollution and the consumers of the natural resources, as generally is the case in the West. In other words, the 'polluters' and 'users' will pay. And, since the polluters and users are, in one way or another, the citizens of the CEE countries, it is they who must assume this responsibility.

NOTES

1. In the following, the terms 'CEE countries' or 'the region' are generally used to cover the following countries (see Figure 1.1): First, the *Czech Republic, Poland, Hungary and Slovakia*. These four countries are also called the Visegrad countries and represent the countries which have advanced most in reforms. Second, the countries of South-Eastern Europe, *Bulgaria, Romania, as well as Croatia, Slovenia, the Former Yugoslav Republic of Macedonia and Albania*. Third, also the three Baltic countries as successor States from the former USSR, *Estonia, Latvia and Lithuania* are covered, as these countries are clearly pursuing an integration into the European Union and other Western institutions or alliances.
2. The Environmental Programme for Europe, a broad environmental strategy which should apply to all ECE countries, is not discussed here. In the process of preparation, the programme's contents and approach were changed substantially several times. One problem was that the programme applies to Western European countries with relatively highly developed environmental policies, as well as to CEE countries in transition, but also to NIS countries, in some of which environmental policy development is virtually at the beginning. All in all, it is an open question whether the programme will bring concrete results.
3. This cross-border facility links PHARE capital investments in CEE countries to parallel actions and projects by neighbouring EU Member States using EU structural funds. This increases investment funds for CEE projects and promotes closer partnership between the involved countries.

REFERENCES

Chabrzyk, G. (1995) Preparation of environmental projects in Bulgaria. In: G. Klaassen and M. Smith (eds), *Financing Environmental Quality in Central and Eastern Europe: An Assessment of International Support*. IIASA, Laxenburg, Austria, October 1995.

COWIconsult (1995) *Case Study of Environmental Expenditure and Investment in Six Selected CEE Countries.* Document prepared for OECD, September 1995, Lyngby, Denmark.

Declaration by the Ministers of Environment of the region of the United Nations Economic Commission for Europe (1995). Document submitted to the Ministerial Conference 'Environment for Europe'. Sofia, October 1995.

Donath, B. (1995) The use of foreign financial assistance in Hungary. In: G. Klaassen and M. Smith (eds), *Financing Environmental Quality in Central and Eastern Europe: An Assessment of International Support.* IIASA, Laxenburg, Austria, October 1995.

EAP Task Force–OECD (1995a) *Report on EAP Task Force Activities 1993–95.* OECD, Paris.

EAP Task Force–OECD (1995b) *EAP Task Force Report to the Sofia Ministerial Conference.* OECD, Paris.

EAP Task Force–OECD (1995c) *A Framework for Developing National Environmental Action Programmes.* OECD, Paris.

EAP Task Force–OECD (1995d) *Progress Report on National Environmental Action Programmes in CEECs and the NIS.* OECD, Paris.

EAP Task Force Secretariat–OECD (1995e) Information prepared for the *Integrated Report on Environmental Financing.* Paris.

European Bank for Reconstruction and Development (1994) *Transition Report 1994.* EBRD, London.

European Bank for Reconstruction and Development (1995) *Transition Report 1995.* EBRD, London.

European Commission (1995a) *PHARE – Progress and Strategy Paper Environment to the Year 2000.* European Commission Directorate General IA, Brussels.

European Commission (1995b) *PHARE – Annual Report 1994.* European Commission Directorate General IA, Brussels.

European Commission (1996) *Preparation of the Associated CEECs for the Approximation of the EU's Environmental Legislation* (Draft). Working document of the European Commission Services, March 1996.

European Parliament (1994) *Report on the PHARE Environment Programme* (doc.A3-361/93). Resolution submitted to the European Parliament on 18 January 1994.

European Union (1995) *White Paper – Preparation of the Associated Countries of Central and Eastern Europe for Integration into the Internal Market of the Union.* Version of 28 April 1995. European Union, Brussels.

Eurostat, European Commission, European Environment Agency, etc. (1995) *Europe's Environment: Statistical Compendium for the Dobris Assessment.* ECSC–EC–EAEC, Luxembourg.

Francis, P. (ed.) (1994) *National Environmental Protection Funds in Central and Eastern Europe.* The Regional Environmental Center (REC), Budapest.

Harvard Institute for International Development: Central and Eastern Europe Environmental Economics and Policy Project and Newly Independent States Environmental Policy Project (1995) Impediments to environmental investments in CEE and NIS. In: G. Klaassen and M. Smith (eds), *Financing Environmental Quality in Central and Eastern Europe: An Assessment of International Support.* IIASA, Laxenburg, Austria, October 1995.

Hungarian State Budget (1994 and 1995, data compiled by Zsuzsa Lehoczki, Budapest, 1996).

Klaassen, G. and Smith, M. (eds) (1995) *Financing Environmental Quality in Central and Eastern Europe: An Assessment of International Support.* IIASA, Laxenburg, Austria, October 1995.

Klarer, J. (ed.) (1994) *Use of Economic Instruments in Environmental Policy in Central and Eastern Europe*. Regional Environmental Centre (REC), Budapest.

Kruszewski, J. (1995) Refining the operating procedures and principles of the National Fund for Environmental Protection and Water Management. Paper prepared for the International Conference Proceedings: *Strengthening Environmental Funds in Economies in Transition*. Warsaw, 1995.

Laurson, P., Melzer, A. and Zylicz, T. (1995) *A Strategy to Enhance Partnerships in Project Financing for Environmental Investments in Central and Eastern Europe*. Document prepared for EBRD, April 1995, London.

Lehoczki, Z. and Peszko, G. (1995) *The St Petersburg Guidelines on Environmental Funds in the Transition to a Market Economy*. OECD, Paris.

Meadows, D. et al. (1973) *The Limits to Growth*. Universe Books, New York.

Nowicki, M. (1995) Poland's experience in financing environmental investments. In: G. Klaassen and M. Smith (eds), *Financing Environmental Quality in Central and Eastern Europe: An Assessment of International Support*. IIASA, Laxenburg, Austria, October 1995.

Organization for Economic Co-operation and Development (1996) *Environmental Performance Reviews – Bulgaria*. OECD/CCET, Paris.

Our Common Future. Report of the World Commission on Environment and Development (WCED), 1987.

PHARE (1995) *Background Paper on Environmental Financing Initiatives*. Document prepared with EBRD as background to the *Integrated Report on Environmental Financing*. Submitted to the Ministerial Conference 'Environment for Europe', Sofia, October 1995.

Polish National Fund for Environmental Protection and Water Management (1996) *Summary Annual Report 1995*, Warsaw.

Popova, T. (1996) Environmental expenditures in Bulgaria. Unpublished paper, Sofia, May 1996.

Project Preparation Committee (1995) *PPC Report*. Document submitted to the Ministerial Conference 'Environment for Europe'. Sofia, October 1995.

Regional Environmental Centre (1994) *Strategic Environmental Issues in Central and Eastern Europe, Volumes I & II*. REC, Budapest.

Regional Environmental Centre (1995) *Status of National Environmental Action Programmes in Central and Eastern Europe*. REC, Budapest.

Regional Environmental Centre (1996) *Approximation of European Union Environmental Legislation – Case Studies of Bulgaria, Czech Republic, Estonia, Hungary, Latvia, Lithuania, Poland, Romania, Slovak Republic and Slovenia*. REC, Budapest.

Stanners, D. and Bourdeau, P., (eds.) (1995) *Europe's Environment – The Dobris Assessment*. European Environment Agency, Copenhagen.

Stepanek, Z. (1995) International financial instruments for environmental investment: Experiences in the Czech Republic. In: G. Klaassen and M. Smith (eds), *Financing Environmental Quality in Central and Eastern Europe: An Assessment of International Support*. IIASA, Laxenburg, Austria, October 1995.

United Nations Economic Commission for Europe (1995a) *Integrated Report on Environmental Financing*. Document prepared with the EAP Task Force Secretariat at OECD and PPC and submitted to the Ministerial Conference 'Environment for Europe'. Sofia, October 1995.

United Nations Economic Commission for Europe (1995b) *An Assessment of the Environmental and Economic Situation in Countries in Transition: Conclusions and Recommendations*. UNECE, Geneva.

United Nations Economic Commission for Europe (1995c) *Industry and Environment – Background Paper by the Secretariat*. UNECE, Geneva.

United Nations Economic Commission for Europe (1995d) *Energy Prices and Taxes in Countries in Transition: An Overview of Recent Developments*. UNECE, Geneva.

World Bank (1992) *World Development Report 1992*. Oxford University Press, New York.

World Bank (1994) *Environmental Action Programme for Central and Eastern Europe*. Abridged version of the document endorsed by the Ministerial Conference 'Environment for Europe'. Lucerne, April 1993.

World Bank (1996) *World Development Report 1996*. Oxford University Press, New York.

CHAPTER 2

Bulgaria

Kristalina Georgieva and Judith Moore

2.1 BACKGROUND

Bulgaria is a small country (111 000 km^2), bestowed with a beautiful and diverse landscape. Before the Second World War, the country's economy was primarily based on small-scale agriculture, which protected the land from many of the long-standing urban and industrial environmental problems of other states in Europe. After the war, under the influence of the Soviet planning model of rapid industrialization and urbanization combined with large-scale agriculture, pollution levels rose dramatically in many regions. The State promoted the development of heavy industries that used highly polluting technologies and demanded large energy and material inputs. Poor city planning placed these industries in the middle of densely settled areas, exposing communities to high levels of ambient pollution. In addition, the agricultural sector adopted an intensive, chemical-dependent model of farming; and like the rest of the economy, transportation and tourism were developed by targeting economies of scale with no consideration of their environmental impacts. As in virtually every other industrialized country at the time, air, water and soil were considered free goods. Natural resources were drastically underpriced, and their consumption generously encouraged. No attempts were made to evaluate the economic losses to the country due to pollution damage or over-consumption of resources.

In the 1970s and 1980s, many market economies changed their development model towards input minimization, energy savings, and pollution abatement and prevention. Bulgaria continued its high resource consumption growth pattern, however, driven by the trade structure and mutual dependencies of the COMECON countries and the artificially low price of natural resources and energy among the trading partners.

When totalitarian Socialism collapsed in 1989 in Bulgaria, the policy of state secrecy over environmental information was abolished. It was

The Environmental Challenge for Central European Economies in Transition.
Edited by J. Klarer and B. Moldan. © 1997 John Wiley & Sons Ltd.

immediately obvious to Bulgarians that environmental conditions and the policies designed to address them were woefully inadequate. As in the other states of Central and Eastern Europe, Bulgaria inherited severe price distortions and subsidies, and state control over resource allocation. All the environmental policy institutions were weak in relationship to the enterprise sector, and they were incapable of enforcing the strict standards and regulations that existed. Although a few critical steps have been taken since economic transition began, such as revision of basic environmental legislation and development of a national environmental strategy, environmental policy development and investments in environmental improvements have been difficult to carry out. Overcoming the heritage of direct central control over society will be a long hard process.

Impact of the political and economic transition of the early 1990s

Economic reforms, including price liberalization, industrial restructuring, and privatization, were initiated in the early 1990s. The economic transition was expected to have a substantial beneficial impact on the environment. Policymakers anticipated that price reforms would discourage inefficient use of energy and other inputs and restructuring would shift production to less polluting industries and technologies. Loosely speaking, this shift appeared to be occurring. Between 1990 and 1994 (when the economy began to rebound) the GDP decreased by over 25%. The greatest fall in output occurred between 1992 and 1993 and was in the industrial sector, particularly among the highly polluting metallurgy, machinery, transport equipment, rubber, and electrical products industries. (Unless otherwise stated, the Bulgarian *Statistical Yearbooks* for each of the respective years are the primary sources of data in this chapter.) Paper and pulp, chemicals, and textiles, also potentially highly polluting, experienced less of a decline (Figure 2.1). Overall, value-added in the industrial sector fell about 40%; and value-added in the agricultural and service sectors fell about 7% and 20% respectively. As a result, during this period, Bulgaria experienced a decrease in industrial air emissions, a decrease in agricultural water pollution, but a relative rise in water pollution by smaller enterprises such as food processing units, many of which are private enterprises that escape pollution supervision.

Since 1994, the economy has improved. The shift in industrial composition has not occurred as expected, or is happening very slowly, and many of the pollution-reduction trends have been reversed – proving that the post-transition environmental improvements were ephemeral. For example, a positive GDP growth rate in 1994 of 1.4%, combined with growing numbers of vehicles, was accompanied by a 4.2% growth in SO_2, 36.6% growth in NO_x, and 31.2% growth in CO emissions. Since 1991, the increase in the number of cars and trucks has caused a doubling in auto emissions' share in air pollution. In 1994, autos were the source of 46.3% of the total NO_x

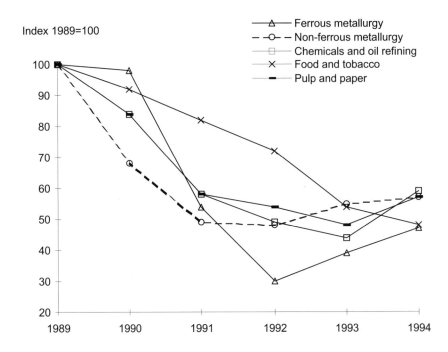

Figure 2.1 Trends in industrial production in Bulgaria, 1989–94. *Note*: Index of production volumes. 1994 data based on preliminary estimates. *Source*: National Statistical Institute

emissions, 43.8% of the hydrocarbons, and 46.7% of the CO. Unfortunately, the strongest industries have been those with the most significant pollution impact, such as ferrous and non-ferrous metals, chemicals, and fertilizers. The impact of economic recovery has, therefore, been particularly negative in a number of 'hot spot' areas, such as Plovdiv-Asenovgrad, Pirdop-Zlatitsa, and Kurdzhali (lead, arsenic, and other metals), and Dimitrovgrad-Maritsa Iztok, Kremikovtsi-Sofia, Pernik, Pleven, and Burgas (particulates and SO_2). The increase in pollution since 1994 confirms the dependence of environmental progress on economic reform and industrial restructuring.

Privatization is proceeding slowly in Bulgaria. By the end of November 1995, the number of state enterprises approved for privatization under the cash privatization scheme reached 1204, but agreements had been finalized with new owners for only 236 of them. Privatization of municipal properties has been more rapid. By January 1996, municipally-owned properties offered for privatization included some 2253 entities, of which about 1110 were transferred to private ownership. A mass privatization plan using vouchers, which was initiated in January 1996, is expected to speed up the process.

Before privatization gets well underway, environmental liabilities must be clarified. Difficulties in monitoring increase the danger that new enterprise owners will blame environmental damage caused by current activities on past activities, and they will be lax about complying with regulations. In addition, clean-up efforts will be needed at some of the more toxic sites. While legislation absolves new owners of any burden related to past pollution, it does not resolve who is responsible or indicate how clean-up should be funded.

The issue of liability is a serious concern of the environmental authorities. All enterprises subject to privatization have been requested to obtain an environmental assessment to determine the current impact of their activities. Depending on the size of the enterprise, this assessment is conducted by either the Regional Environmental Inspectorates (REIs) or by the Ministry of Environment (MOE). So far, the Ministry has initiated more that 350 environmental assessments. Due to staff shortages and the large number of applicants, however, some of these assessments are incomplete, and they can be challenged or ignored. Government regulation requires that environmental audits be conducted in enterprises destined for privatization. A project promoting environmental audits is currently underway in the large enterprises known to be key point sources of pollution, such as large refineries, heating and power plants, and smelters. These audits should provide better information on the current conditions at those sites. Environmental audits have already been conducted at some sites, resulting in significant improvements in environmental conditions at low or no cost.

As long as state ownership of the major industries continues to exist, environmental regulation and clean-up will be hampered, as the regulator and the subject of regulation are, for all practical purposes, one and the same. In the past, when the primary concerns of the state apparatus were in compliance with ideological requirements and production targets, no effective environmental policy could be initiated. Likewise, an effective industrial pollution policy will continue to elude the government so long as pressures from widespread unemployment, short-term economic goals, severe governmental budget limitations, and reliance on old ways of doing business constrain market reforms and governmental enforcement of environmental standards.

Past and current state of the environment

Air Quality

The economic recession in the early 1990s corresponded with general improvements in air quality (Table 2.1). As noted earlier, most of these improvements were the result of production cutbacks and industrial shifts. Some were directly related to environmental control measures that were

Table 2.1   Air quality in selected 'hot spots' in Bulgaria (annual averages, μg m^{-3})

City	Particulates		Sulphur dioxide		Lead	
	1990	1993	1990	1993	1990	1993
Asenovgrad	344	180	252	113	2.33	0.52
Burgas	88	101	36	34	–	0.2
Dimitrovgrad	380	216	71	86	0.9	–
Galabovo	125	224	72	117	0.11	0.1
Kremikovtsi	118	126	37	26	0.51	0.39
Kurdzhali	327	169	89	130	1.1	1.1
Pernik	179	387	122	44	0.48	–
Pirdop	453	208	330	206	–	–
Pleven	309	180	8	21	0.3	0.3
Plovdiv	485	360	35	88	1.5	0.7
Rousse	157	182	28	19	–	–
Sofia-Drujba	245	165	36	23	0.27	0.26
Svishtov	509	103	18	96	–	0.1
Varna	432	545	29	31	–	0.2
Veliko/Turnovo	491	257	33	27	0.47	0.4
Zlatitsa	432	185	232	345	–	–

Adapted from Table 4 – Air Quality Trends by Urban Area, *Bulgaria Environmental Strategy Study Update and Follow-Up*, 30 December 1994.

instituted since 1990 (Box 2.1). In particular, exposures to lead dust, particulates, and sulphur dioxide in the Asenovgrad and Kurdzhali regions have fallen dramatically since low-cost pollution control measures were instituted in two smelters. Figure 2.2 illustrates the impact of control measures on lead emissions from the Kurdzhali smelter. Particulates in Galabovo, Pernik, and Rousse have increased, however, probably because of an increase in the use of low quality fuels in power/heating plants and in homes. Consistent data are not available for all towns, so there may be some, such as Devnia, that should be included in the 'hot spots' category but are not. Currently, Asenovgrad-Plovdiv, Dimitrovgrad-Galabovo, Kurdzhali, Pernik, and Pirdop-Zlatitsa are the industrial areas with the greatest air quality problems. Air pollution is an increasing problem in Sofia as well, mainly due to vehicular congestion.

In 1994, air pollutants above the permissible levels were registered in almost all areas that had been classified as 'hot spots' in 1991. Average annual concentrations in industrial and urban areas were significantly above the norms for particulates and SO$_2$. Figures 2.3 and 2.4 show data for these two pollutants in the ten worst cities. (The bar charts represent data averages for specific monitoring points and do not reflect regional air shed averages.) Lead concentrations were generally within the norms, although traffic-related airborne lead has increased.

Box 2.1   Controlling lead emissions

In the late 1980s, 90% of the atmospheric lead emissions in Bulgaria were from large stationary sources. The remaining 10% was generated primarily by motor vehicles. In the last few years, lead emissions have been reduced due to declining industrial production, changes in the lead content of fuels, and emissions' controls imposed at the sources. Nevertheless, lead levels in the environment continue to be a source of concern. Poor land-use planning placed residential areas in unacceptably close proximity to highly polluting plants, and lead smelters were built in the middle of some of the richest farmlands in the country. In studies conducted in the late 1980s in Sofia and other key cities, excess levels of lead were found in foods in all seasons. Children living near smelters had average blood levels that were considerably above safe limits, and it was unknown whether the lead came from food, soil, household dust, aerosols, or other sources.

Emission control measures instituted since 1991 at several 'hot spots' have successfully reduced lead exposures in populations surrounding the plants. As a result of controls imposed at the smelter in Kurdzhali, for example, the industry share of lead emissions dropped from 90% in 1990 to about 17% in 1993, despite a rise in lead production at the smelter. After 1993, a comparative analysis of the share of airborne lead pollution from various sources in the region showed that 73.4% of the ambient lead came from motor vehicles, 17.3% from lead smelters, and 9.3% from a power plant.

Vehicle-related lead emissions are expected to rise. Although the lead content of gasoline has declined from about 0.25 g^{-1} l in 1987 to about 0.15 g^{-1} l, the number of cars on the road increased from about 1.85 million in 1990 to 1.97 million in 1993, and it is still growing. Many secondhand cars that use leaded gasoline have been imported from Western Europe. As of July 1995, about 83% of the drivers in Sofia used leaded gasoline in their cars.

The Ministry of Environment, in partnership with industrial, transport, health and finance authorities, and NGOs, has initiated the development of a national strategy for phasing out leaded gasoline. The strategy will introduce a comprehensive package of policy incentives for switching to unleaded gasoline by the beginning of the twenty-first century. Neftochim, the biggest Bulgarian oil refinery, is exploring potential investment sources to finance a refinery upgrade to increase unleaded gasoline production. In addition, a new import tax structure is being established that will discourage the importation of older automobiles that use leaded gasoline.

Source: *Lead Exposure and Health in Central and Eastern Europe*, 1995. Magda Lovei and Barry S. Levy (eds), The World Bank.

Industrial recovery should be combined with measures to reduce emissions from big point sources and to improve control over small and non-point sources. Reducing the high levels of particulates is especially critical, as they are a serious health hazard especially when combined with SO_2. To achieve long-lasting improvements, the energy sector in Bulgaria must be reformed and modernized, and a coal to natural gas conversion achieved. Currently, there is no municipal gas distribution, although a number of cities would like to pursue gasification, and there is interest among local gas companies, foreign firms and donors. US AID and the World Bank are supporting efforts to

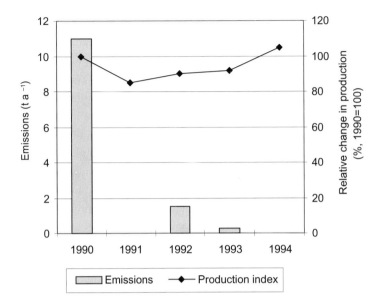

Figure 2.2   Lead production and lead emissions by the lead smelter in Kurdzhali, Bulgaria, 1990–94

improve the regulatory regime for municipal gasification through several pilot projects. If these and other programmes successfully reduce air pollution in urban and industrial areas, they should yield significant health and economic benefits.

Water and Soil Quality, Waste Management

Water is a relatively scarce resource in the country with a per capita endowment of 2400 m³, less than half the European average of 5000 m³. Seasonal and year-round imbalances in water supply persist in many cities and villages. In 1994–95 a critical water shortage occurred in the capital creating serious health risks. Nevertheless, drinking water quality has remained good overall, although there is nitrate pollution in some regions, and water pollution from industrial effluents, feedlot wastes, and municipal sewage can be severe in localized areas. Industrial activities, mining and construction have been commonly conducted without regard to environmental impacts and have been major local contributors to soil degradation and ground and surface water pollution. Many municipal waste sites have also been operated as uncontrolled dumps, with toxic and hazardous wastes mixed with domestic wastes. In these dumps, the potential exists for hazardous chemicals leaching into the groundwater.

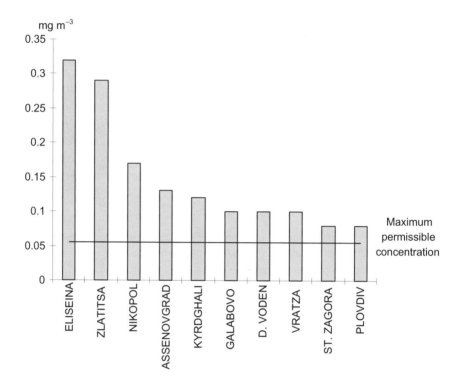

Figure 2.3   Points with highest average annual concentrations of SO_2 in Bulgaria in 1994. *Source: Annual Report on the State of the Environment in 1994 (Green Book)* (1995)

Although there are a number of monitoring sites for water quality throughout the country and excellent equipment has been provided by donors such as PHARE, there remain large variances in the data making it difficult to summarize trends. In general, however, the data indicate that nitrates and suspended solids have declined significantly since 1990. Nitrate levels, in particular, have fallen dramatically and are now well below EU standards most likely due to higher agricultural input prices and a large (about 40%) decline in the size of livestock herds. The decline in industrial activity has benefited water quality in the Jantra, Maritsa and Tundja River Basins, but BOD5 and suspended solids are still high in the Rousenksy Lom, Kamchija, Osam and Provadijska Rivers and in Mandrensko Lake. Information about heavy metal contamination of rivers and groundwater in the areas surrounding metallurgical plants is incomplete.

Wastewater treatment is a major concern for the environmental authorities in the country. Currently three main instruments for control of water pollution

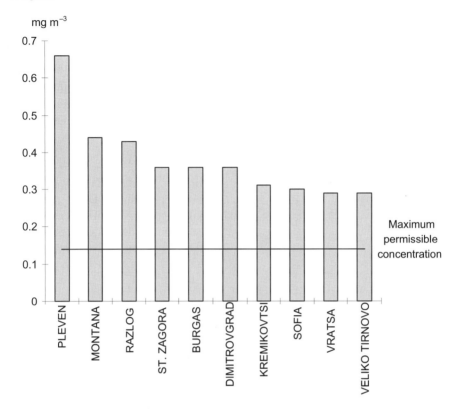

Figure 2.4 Points with highest average annual concentrations of particulates in Bulgaria in 1994. *Source*: *Annual Report on the State of the Environment in 1994 (Green Book)* (1995)

are applied: water quality standards, discharge permits and fines for excess pollution. A system of pollution charges for discharge within the permissible limits and penalties for discharge above the limits has been designed but not yet implemented. Monitoring ambient water quality and effluent quality for point sources is sufficient and provides a solid basis for setting priorities for integrated river basin management.

Both industries and municipalities are required to discharge water within the water quality standards of the receiving body of water (sewage systems or rivers). All industries potentially discharging wastewater above these standards are under a legal obligation to have adequate treatment facilities. There are three main weaknesses in the system of industrial wastewater control: (i) although plants may have treatment facilities, they are often inadequate to mitigate the effluents emitted or they are not properly functioning; (ii) the systems may be adequate and functioning but they are in poor repair; and (iii) some plant facilities were designed to discharge into a municipal sewage system that is connected to a municipal sewerage plant, but the municipal plant either is not working or has not been built yet.

Almost all major cities have functioning sewerage treatment plants although they are experiencing serious difficulties in operating at full capacity. Currently, there are about 30 municipal wastewater treatment plants at various stages of construction with about 4.5 billion leva (US$62 million) invested in them and limited funds available for completion. Municipalities are dependent on state budget transfers and funding from the National Environmental Protection Fund to start or complete the plants. Until now the transfers were spread so widely that no single plant of those under construction could be completed. The MOE is now trying to prioritize funding based on which plants will have the highest impact on water quality. Until recently, wastewater treatment has been regarded as a free good by municipalities with much of the planned plant capacity far higher than needed. User fees do not cover operations now, much less construction costs, but these fees will be raised. Until fees are related to discharge contents, enterprises will have little incentive to reduce effluent loads. Under pressure of changing economic conditions, and with support from the current World Bank Water Sector Loan, municipalities must assume financial responsibility for plant operations. It is likely that charges and planned capacity will then become more rational.

Soil quality in many regions of the country is closely related to the water quality concerns in those regions. Wind and water erosion affects about 30% of Bulgaria's arable lands. Poorly designed gravity irrigation systems are causing erosion on about 38% of the irrigated lands with associated siltation occurring in adjacent streams; irrigating with severely polluted water has contaminated soils in broad areas; lack of proper drainage in other areas has caused soil and water salination. Some sites have been polluted by mine tailings and other wastes. In the Vidin region, waters are contaminated by mines in the former Yugoslavia. Aerosol emissions from metallurgical plants in Elisseina, Kurdzhali, Kremikovtsi, Pernik, Plovdiv, and Pirdop have severely contaminated local soils with lead, zinc, cadmium, arsenic, and copper. Some of this heavy metal contamination has occurred on what used to be prime agricultural land.

Regarding solid wastes, all cities have municipal collection and many industries maintain their own landfills. Due to severe financial constraints of the municipalities in the last couple of years, waste collection and disposal services have deteriorated. The economic difficulties of transition have also created a positive impact on recycling. Almost all waste that is currently economically and technically suitable for recycling is collected from both waste sites and industrial backyards but often by informal collectors. This practice creates the risk of serious health hazards during the process of collection because of the unsorted nature of the dumps.

For hazardous waste, the Ministry of Environment has forged a basic legal and regulatory framework that defines what constitutes hazardous waste, compels the waste generators to report information on generation and management, and requires that releases from waste management activities are safe for the environment and for human health. The MOE plans to deal with hazardous waste are supported by a US sponsored study. According to the study, in 1994 hazardous waste totalled approximately 14 million tonnes, of which about 13 million tonnes were mineral processing wastes. As in most countries, both generation and management are skewed to the largest facilities – 50 facilities generate about 90% of the waste and 30 facilities manage about 90% of the waste. Metallurgical and petrochemical plants dominate both generation and management. The primary management practice is disposal in landfills. There are no treatment requirements for waste, nor are there monitoring requirements for groundwater or other environmental media, so the extent and risk of releases from these facilities is largely unknown. There are few incentives for facilities to minimize the volume of waste generated or for managers to safely contain wastes. Enforcement and fines are too low to stimulate changes in the behaviour of facility managers.

The recommendations of the study address both pollution control (i.e. controlling releases once wastes are generated) and pollution prevention (i.e. minimizing the volume or toxicity of waste generated in the first instance). Seizing on the opportunity to solve a large share of the problem by focusing on a small number of facilities, the study's five-year Action Plan calls for facility permitting, environmental monitoring, and use of economic incentives to improve the environment while improving efficiency. Bulgaria is also developing a set of project proposals that include building treatment and disposal capacity and demonstrating pollution prevention techniques.

Protected Areas

Because of its highly varied topography and climate, Bulgaria is one of the most biologically diverse countries in Europe. The country supports at least 94 species of mammals, 383 species of birds, 52 species of reptiles and amphibians, 207 Black Sea and freshwater fish, and thousands of invertebrate species, plants and fungi. Many of these species have economic uses that are widely recognized and may provide significant economic benefits if they are properly preserved. All ecosystems are threatened by human encroachment, however, from the high mountains to the benthic communities of the Black Sea. Both point and non-point source pollution are ubiquitous in the land-scape, and legal and illegal harvesting of fungi, medicinal plants, fish, birds, and mammals threaten many species. Land restitution to former owners, which is currently underway, is both an opportunity and a potential setback for biodiversity conservation efforts. Citizens and local governments need to be better informed on the need for ecosystem conservation and encouraged to practise conservative and restorative land management.

Table 2.2   Protected areas (km^2) in Bulgaria, 1977–95

	1977	1985	1990	1992	1994	As % of land area in 1995
Strict Reserves	234	547	610	770	807	0.7
National Parks	447	709	1136	2493	2462	2.3
Natural Monuments	146	220	233	228	230	0.2
Protected Historical Sites	236	290	293	342	342	0.3
TOTAL	1063	1766	2272	3822	3841	3.5

Source: *Statistical Yearbook of Bulgaria*; Environment '94, National Statistical Institute.

Bulgaria has a well-defined network of protected areas, including 17 biosphere reserves. Many of these protected areas are a legacy of Todor Zhivkov's long rule and state militarism. Zhivkov's love of hunting and luxury residences led to exclusive protection of his preferred hunting areas and the set-aside of a number of country estates around the country. In addition, for strategic reasons, zones near the borders with Romania, the former Yugoslavia, Turkey, and Greece were deliberately undeveloped and residents were relocated. These policies preserved large areas in these regions in a relatively pristine condition. As of 1995, a total of 4.5% of the country has been set aside (see Table 2.2). These sites are now ideal for quick introduction of environmentally-friendly businesses such as small-scale ecotourism. Further development must be carefully considered, however. The primary threat to these and all protected areas has been and remains extensive, poorly-planned development for industrial, recreational, and other purposes.

In addition to protected areas, Bulgaria has extensive forests covering about one-third of the land area. All forests are officially owned by the State, although privatization of some areas is proceeding. The Government expects to return about 57% of the forest lands to municipal management, and about 19% to private owners. A law on forest ownership has been drafted and will be submitted to the National Assembly at the same time as the new Forest Law.

The forests are relatively young, due to extensive over-harvesting and reforestation between the 1950s and the 1970s. The most widespread stands are between 21 and 40 years old. Although forests in Bulgaria are generally less affected by pollution than other Central European forests, soil erosion, acidification, waterlogging, salinity, and heavy metals contamination have severely damaged at least one-fifth of the forest area. In addition, many of the forest stands are physiologically unstable, because they are composed of species that were planted outside of their natural range or they are young, dense monocultures. These problems, combined with chronic exposure to air pollution, are resulting in a general decline in forest health. Remedies, such as replacement and thinning of the stands, must be financed by revenues obtained from harvests. The lack of markets for the existing, low-value wood is hampering efforts to improve forest stands and reform forest management.

Transboundary pollution issues

In the past, transboundary pollution was a strongly motivating political issue in Bulgaria. During the late 1980s, public concern about the effect of air pollution crossing the border between Romania and Bulgaria spurred the making of a provocative documentary called *Breathe* that broke the iron silence on environmental issues imposed by the Zhivkov regime. The environmental NGO Ecoglasnost and other organizers used public concern over the condition of the environment to focus political opposition to the Government. They ultimately precipitated the collapse of the Zhivkov Government and the beginning of economic and political transformation. Transboundary pollution is still a vital issue in the country focusing primarily on air and water pollution problems between Bulgaria and Romania but also including shared concerns with the former Yugoslavia, and the Black Sea nations.

Transboundary Air Pollution

Overall, Bulgaria is a net exporter of sulphuric oxides (SO_x), and a net importer of nitrous oxides (NO_x) (see Tables 2.3 and 2.4). In particular, sulphur dioxide and ammonia blow across the border into Bulgaria from large industrial enterprises in Romania. The biggest source of air pollution along the boundary is the S.C. Turnu S.A. fertilizer plant in Turnu Magurele, Romania. Emissions from the plant affect Nikopol, Bulgaria, where peak air pollution levels sometimes exceed the measuring capabilities of the instruments used to record the data. In addition, transboundary dust or particulates, caused mainly by industrial emissions, heating and power plants, and household heating, are a problem in all of the border cities for which data are available. It should be noted that data are not entirely reliable. For example, the method of sampling for dust includes total dust, not the fraction below 10 microns, which is the critical fraction for assessing human health risks. As in other parts of the country, air quality in the transboundary region has significantly improved since economic transition began, mostly due to reductions in industrial output in both countries. Table 2.5 summarizes the ambient air quality problems as reported in inception reports provided to a CEC PHARE fact-finding mission.

Transboundary Water Pollution

The Danube River Basin and the Black Sea are the focus of most of the transboundary water pollution concerns in Bulgaria. Bulgaria has joined

Table 2.3   Primary destinations of Bulgarian SO_x and NO_x emissions, 1993

Recipient region	Sulphur oxides (SO_x)		Nitrous oxides (NO_x)	
	Tonnes $(\times 100)$	%	Tonnes $(\times 100)$	%
Bulgaria	1240	40	37	14
Greece	248	8	24	9
Romania	124	4	19	7
Russia	124	4	24	9
Turkey	186	6	19	7
Ukraine	93	3	16	6
Mediterranean	403	13	45	17
Black Sea	372	12	21	8
Other	310	10	61	23
TOTAL	3100	100	266	100

Source: OECD data.

Table 2.4   Primary sources of acid deposition in Bulgaria, 1993

Country of origin	Sulphur oxides (SO_x)		Nitrous oxides (NO_x)	
	Tonnes $(\times 100)$	%	Tonnes $(\times 100)$	%
Bulgaria	1224	56	38	11
Romania	251	11	59	17
Serbia	138	6	19	6
Ukraine	66	3	18	5
Hungary	62	3	11	3
Germany	53	2	38	11
Greece	16	1	22	6
Turkey	6	1	2	1
Other	379	17	136	40
TOTAL	2195	100	343	100

Source: OECD data.

international efforts to control and correct problems in both watersheds and has signed international agreements to harmonize standards and share resources to protect these waters (Box 2.3).

Danube River   The Danube River is 2857 km long in total and passes through or between 17 countries. The lower end of the Danube marks the border between Bulgaria and Romania. The river supplies drinking and irrigation water and supports industry, transport, power generation, and

Table 2.5 Ambient air quality problems in the Bulgaria/Romania transboundary region, 1994

Country/City	SO$_2$	NH$_3$	H$_2$S	HCl	Phenol	Dust
Bulgaria						
Silistra			×		×	×
Rousse			×	×		×
Nikopol	×	×	×			
Romania						
Calarasi						
Giurgiu						

Source: Adapted from the CEC PHARE Final Report on *Transboundary Pollution Romania – Bulgaria*.

sewage disposal throughout its course. High nutrient and sediment loads, toxic wastes, contamination with microbial and oxygen-depleting substances, water diversions and changes in the river patterns all pose serious threats to human and natural communities along its length. Bulgaria contributes only about 4% of the total pollution in the Danube but has joined international efforts to improve the quality of the waters. The Government has signed the *Convention on Cooperation for the Protection and Sustainable Use of the Danube River Basin*, as well as participated in the development of the Environmental Action Programme for the Danube River Basin. The *Action Plan* is built on several principles: pollution should be prevented in the first place, rather than mitigated later (called the Precautionary Principle); best available techniques and best environmental practice should be applied in industrial development planning and pollution control; market principles, including the 'polluter pays' principle, are key tools in the development of policy measures; and countries in the river basin must share information, harmonize standards, and generally co-operate on a regional level.

A number of problems on the Danube that require particular attention have been identified in the *Strategic Action Plan for the Danube River Basin: 1995–2005* as having their source in Bulgaria. It is important to reiterate, however, that cleaning up the Danube is a multinational problem with pollution sources spread from Germany to the Black Sea coast. The problems emanating from Bulgaria are not necessarily the most critical ones for the Danube Basin, nor the most pressing problems in Bulgaria itself. Nevertheless, municipal and industrial wastes from a number of sites are of concern (Box 2.2) particularly effluents from the Sviloza Chemical Plant, which produces cellulose and viscose, and the Prista Tannery and Leather Products firm.

The Black Sea The Black Sea is one of the most polluted bodies of water in the world. About 165 million people contribute directly or indirectly to its pollution problems, and control of eutrophication, siltation, and chemical

Box 2.2   Pollution in the Danube River Basin, Bulgaria

River Basin	Site	Problem	Action Required
Danube	Bulgarian Bank of the Danube	Erosion and abrasion on the Bulgarian Bank due to 'Iron Gate I and II' upstream in Serbia and Romania. Natural river flow strongly disturbed.	Improvement of management of upstream reservoirs and power generation stations. Rehabilitation of the Bulgarian Bank.
Yantra	Gorna Orahovitza	Effluents from municipal wastewater treatment plant and sugar and alcohol factory (BOD and SSM).	Construction of a new wastewater treatment plant, construction of treatment facilities at the factory; technological modernization of the plants.
Yantra	Sevlievo	Effluents from the municipal wastewater treatment plant (BOD, SSM).	Construction of a new wastewater plant.
Yantra/ Rossitza	Sevlievo 'Sevko'	Effluents from local tannery (BOD, SSM and solids).	Construction of a new factory treatment plant.
Osam	Lovetch	Effluents from municipal wastewater treatment plant (BOD, SSM).	Existing plant needs to be rehabilitated and expanded.
Ogosta	Montana	Effluents from municipal wastewater treatment plant.	Construction of a new wastewater treatment plant.
Vit	Pleven	Untreated effluents from municipal wastewater treatment plant.	Existing plant needs to be rehabilitated and expanded.
Russenski Lom/Beli Lom	Razgrad 'Antibiotic'	Effluents from the chemical and pharmaceutical plant (BOD, SSM).	Complete construction of the onsite treatment plant.
Iskar	Sofia	Effluents from the municipal wastewater treatment plant and untreated effluents (BOD, SSM).	Existing plant needs to be rehabilitated and expanded, and main collectors finalized.
Iskar	Sofia–Iskar Reservoir	Severe water shortages, water quality problems.	Controls on water quality protection, and watershed landuse implemented; existing wastewater treatment plant expanded.
Osam/Beli Osam	Troyan	Effluents from the municipal wastewater treatment plant (BOD, SSM).	Construction of new wastewater treatment plant and sewer collectors.
Ogota/Leva Betaine	Vratza	Effluents from the municipal wastewater treatment plant.	Existing plant needs to be rehabilitated and expanded.
Ogosta/ Dubnica	Vratza 'Himco'	Effluents from the local fertilizer plant (BOD, SSM, TDS, NH_3).	Construction of an onsite treatment plant, and technological modernization.

Source: Adapted from the *Strategic Action Plan for the Danube River Basin: 1995–2005.*

contamination in the Sea can only be accomplished through a concerted international effort. Bulgaria has signed the multinational Convention on the Protection of the Black Sea from Pollution, and participated with other coastal countries in the development of the Black Sea Action Plan. On the national level, better coastal zone management, pollution prevention and mitigation measures, and strengthened enforcement of environmental regulations are needed. Bulgaria has initiated a Coastal Zone Management (CZM) programme with financial assistance from an Institutional Development Facility grant from the World Bank. Rules and standards for the Black Sea Coastal Region were adopted by the Ministerial Council in 1993, and in 1994 a new CZM office was created with regional offices in Varna and Burgas – cities where water recreation and eating fish from the sea are considered unsafe. On the international level, co-operative efforts need to be strengthened to harmonize standards and enforcement, reduce over-exploitation of the Sea's resources, and control inappropriate development on the coasts. Collective efforts to control pollution in the Danube River basin will help reduce high nutrient and sediment loads entering the Sea, as well as contamination by toxic chemicals.

Kozloduy Atomic Power Plant

The Kozloduy Atomic Power Plant, located near the Danube River, is a major issue of dispute with neighbouring countries and the European Union. Reactor units 1–4 at this plant are old Soviet-designed VVER 440 Model 230s, and are considered unsafe by many experts. There have not been any major environmental problems with these units so far, but Bulgaria has agreed with the Nuclear Safety Account of the EBRD to shut down the reactors when a feasible alternative power supply is available. Currently, about 40% of the country's electrical power is generated by this plant. The relative price advantage of domestically-produced energy, as well as concern about security of energy supplies, has pushed Bulgaria to focus even more on nuclear power today than before 1990. Under pressure to reduce SO_2 emissions from the energy sector, even the Ministry of the Environment appears to support a move towards more nuclear energy in the future. The Government's nuclear safety programme calls for enhancing and upgrading the safety systems at all reactors and developing a licensing process in accordance with international standards, with the help of the European Bank of Reconstruction and Development and the International Atomic Energy Agency. Shut-down of the current reactors is not considered an option at this time by the Government. Construction plans for a second nuclear plant, discontinued because of financial constraints and pressure from environmental advocates, are being actively discussed and only the high price tag (approximately $1.2 billion) is hindering a government commitment to build it.

International Co-operation

Bulgaria recognizes the critical importance of international co-operation on environmental management and has signed around 30 international conventions directly and indirectly dealing with environmental issues (although not all are ratified and published yet). Some of the key agreements are listed in Box 2.3. In particular, meetings are held several times a year between high level representatives of the Bulgarian and Romanian Ministries of the Environment, in addition to frequent contacts among municipal representatives from each country, to discuss transboundary pollution. In 1993, the Romanian and Bulgarian Ministers of Environment signed a Memorandum of Understanding expressing mutual willingness to co-operate on transboundary issues. Pollution standards and data collection differ in the two countries, however, complicating negotiations over transboundary pollution. Bulgaria has signed a European Association Agreement in 1995 regarding analytical methods and standards and will be harmonizing its data collection and standards with EU practices in the coming years. As Romania has also signed an Association Agreement, some of the problems between the two countries regarding disputes over data collection and policy standards should be eased.

Bulgaria has an open and proactive policy of international co-operation with bilateral and multilateral partners. The Ministry of Environment is well respected by the international community. One of the symbols of recognition for the Bulgarian achievements in reforming the country's environmental policy and institutions was the selection of Bulgaria as the host country for the third Environment for Europe conference. Ministers and senior officials from more than 50 countries, as well as representatives from international environmental, development, and donor organizations, came to Sofia in October 1995. The conference was generally considered a success by the international participants. In preparation for the conference, the Bulgarian Ministry of Environment took the lead in promoting consultations among the Central and Eastern European countries on initiatives for accelerated implementation of the Environmental Action Programme, adopted in 1993 in Lucerne.

2.2 LEGAL AND ADMINISTRATIVE FRAMEWORK

During the first years of the democratic transition, the Government regarded environmental policy reform as a high priority. Environmental NGOs and other environmental advocates were well represented in the first democratically elected National Assembly; and the relatively 'green' Parliament, energized by the euphoria of the first two years of political transformation, produced ambitious programme platforms. Very good comprehensive environmental legislation was initiated, and environmental institutions were generally strengthened. As the economic reforms advanced, and the country

Box 2.3   Regional programmes and international agreements on the environment –
signed/ratified by Bulgaria

Convention	Date	Purpose
Environmental Programme for the Danube River Basin	1991	To develop a basin-wide programme for environmental monitoring, institutional capacity-building and emergency response.
Bilateral Convention on Environmental Co-operation between Bulgaria and Romania	1992	To co-ordinate efforts to reduce pollution loads in the Danube by jointly assessing transboundary pollution levels, harmonizing standards and data-collection, co-operate on response to industrial accidents, and jointly review development along the border through an intergovernmental committee.
Convention on the Protection and Use of Transboundary Watercourses and International Lakes	1992	To provide a framework for regional co-operation on transboundary problems involving surface and groundwaters.
Convention on Biological Diversity	1992	To conserve biological diversity, share equitably the benefits of genetic resources, and to promote the sustainable use of all natural resources.
Declaration on the Protection of the Black Sea	1993	To protect and restore the Black Sea through the harmonization of standards among the bordering states, integrated coastal zone management, environmental impact assessment, and control of point and non-point source pollution.
Environmental Action Programme for Central and Eastern Europe	1993	To co-ordinate national and international efforts to integrate environmental considerations into the process of economic reconstruction, build institutional capacity in environmental management, prevent or mitigate transboundary pollution issues, and provide immediate assistance to alleviate environmental 'hot spots'.
The Black Sea Environmental Programme	1993	To establish a common policy framework among the bordering countries on principal approaches, goals and priorities for regional action.
Convention on Co-operation for the Protection and Sustainable Use of the Danube River Basin	1994	To conserve and rationalize the use of surface and groundwaters in the Basin, control hazards, and reduce the pollution loads entering the Black Sea.
European Association Agreement	1995	To harmonize policies, laws and standards, including on environmental issues, between the EU, Bulgaria, and other CEE countries.

was forced to confront serious social and economic difficulties, public attention turned from environmental issues to the pressing demands of day-to-day
survival. The task of reforming all of the nation's political and economic
institutions overwhelmed the well-conceived efforts to reform environmental
management.

Legislative developments since 1990

The initial wave of public zeal produced a new comprehensive Environmental
Protection Law in 1991. Since the passage of this law, a package of comple-
mentary laws has been drafted, but little, if any, serious environmental legislation has been adopted. In the national assemblies that followed the first
environmentally inclined Parliament, enthusiasm dwindled markedly. Nevertheless, the 1991 law provided a solid framework for constructing a new
environmental policy and regulatory regime in Bulgaria. It was based on three
key principles: (i) the 'polluter pays' principle, providing the legislative basis
for emission standards, pollution taxes and fees; (ii) the pollution prevention
and precautionary principle, which emphasizes that legislation should be
designed to avoid pollution before it occurs rather than promote pollution
mitigation or clean-up activities; and (iii) the public right to know principle,
which ordains open environmental decision-making processes and free access
to information on the state of the environment. In addition, the law defines
the scope of responsibilities and enforcement powers of the national and local
government agencies and requires the integration of environmental considerations into the activities of other agencies through such instruments as
Environmental Impact Assessments (EIAs).

The Environmental Protection Law was passed with the understanding that
the specific management practices, norms, permit regimes, pollution charges
and other regulatory instruments would be detailed in media-specific laws and
ministerial council ordinances. Since enactment of the law, a number of these
supporting laws and regulations have been prepared addressing clean air,
water resources management, waste management, noise management, nature
conservation, natural resource management, marine environmental protection,
forestry, biodiversity, medicinal plants, game, protected areas, and underground resources. Some of these draft laws have received ministerial council
approval and are in Parliament; others are in an advanced stage awaiting
approval. None has been adopted by the Assembly.

Confronted with an economy that is rebounding in the most polluting
industrial sectors, the general public is once again applying pressure on the
Government to address unresolved environmental concerns. In the beginning
of 1996, the draft clean air law began to progress through Parliament by the
hard effort of MOE. This law will replace the general atmospheric emissions
standards that previously existed with specific site or air-shed emissions limits
to facilitate monitoring and the permitting process. The Government

encourages the continued growth of industries, but enterprises must now do so within the constraints of the 'polluter pays' principle and under pressure to harmonize with European Union standards.

Progress has also been made by the national and regional environmental authorities in levying and collecting pollution fees and fines from enterprises. The 1991 Environmental Protection Law significantly raised pollution fines. Since then, the Ministry of Environment has been working on a number of instruments for implementing the 'polluter pays' principle. Some of them are already realized, including import tax relief for environmental technology, an environmental tax on secondhand cars, and an environmental component of a liquid fossil fuels tax. The Privatization Law requires that 5% of all the privatization revenues be placed in the National Environmental Protection Fund, discussed below.

Parliamentary approval of the Law on Liquid Fuels Tax was particularly significant. The tax was accepted despite the opposition of a strong energy lobby, which was defeated by the concerted efforts of the transport and environmental authorities in dialogue with the Finance Ministry. Part of the receipts from the tax will go into the National Environmental Protection Fund, doubling its expected revenues. Passage of the tax was facilitated by legislators' desire to harmonize with the European Union, which heavily taxes fuels. Overall, the tax is estimated to have an inflationary impact on the economy of 1.0 to 2.5%.

The introduction of economic instruments for environmental protection in Bulgaria, as elsewhere, requires significant effort including extensive negotiations with the affected parties. Agreements are not always easy to reach. For example, the MOE has been actively promoting water pollution charges for the last three years. While there is a general agreement among the concerned government agencies (MOE, the Ministry of Territorial Development and Construction, the National Water Council) that the proposed charge system should be tested on a pilot basis, no agreement to act has been reached. At the same time, raising funds to complete wastewater treatment plants is considered a high priority. Water pollution charges could provide revenues for this purpose.

Table 2.6 lists the primary environmental laws and regulations passed since 1990. From a policy reform perspective, they represent two key achievements: (i) the establishment of an environmental policy framework under the Environmental Protection Law that incorporates market incentives, such as the 'polluter pays' principle; and (ii) the development of regulations governing Environmental Impact Assessments (EIAs).

Although the Environmental Protection Law is comprehensive and well designed, there are two weaknesses in it: (i) as discussed above, it must be supported by additional legislation that does not exist at this time; and (ii) it is subject to amendments that may alter its intent. Regulations governing EIAs, passed in 1992, detail the criteria and procedures to be followed by agencies

Table 2.6   Major environmental legislation in Bulgaria, 1990–95

Date	Legislation
1990	Regulation on Protection Against Accidents with Hazardous Chemicals
1991	Standards for Admissible Emissions of Harmful Substances Discharged in the Ambient Air
1991	Environmental Protection Law
1991	Memo on Maximum Permissible Level of Emissions into Air Adopted by the Ministry of Environment
1992	Regulation on Collection, Spending, and Control of the Financial Resources of the Environmental Protection Funds
1992	Regulation on Environmental Impact Assessments
1993	Decree of the Council of Ministers for the Collection, Transportation, Storage and Neutralization of Hazardous Waste
1993	Regulation on Hygienic Requirements for Health Protection of the Urban Environment
1993	Regulation of the Council of Ministers on the Procedures for Assessing and Imposing Sanctions for Environmental Damage or Pollution Beyond Permissible Limits
1995	Amendment to the Environmental Protection Law
1996	Law on Liquid Fossil Fuels Tax

and enterprises. By the end of 1994, 1000 EIAs had been completed, including 41 large investment projects involving the energy, transport, and wastewater treatment sectors among others. The PHARE Programme has funded training in assessment procedures for environmental professionals in the Ministry and Regional Environmental Inspectorates, and has facilitated staff education in the industrial enterprises. A related programme for environmental audits is also well underway, funded with external assistance. Starting in 1996, all of the big polluting enterprises will be audited.

In 1995, an amendment gave MOE emergency powers to approve completion of projects for which EIAs have not been conducted thus avoiding the mandated public participation procedures. Passage of this amendment was driven by a severe water shortage in Sofia. The newly elected Socialist Government was subjected to strong public pressure to increase the water supply for the city by completing a water transfer project for which no EIA had been conducted. Parliament amended the law to speed the construction process. Unfortunately, what exactly entails an emergency in the EIA context was not made clear in the amendment, and the MOE, as well as the Parliamentary Committee on Environment (chaired by a representative of Ecoglasnost Political Club), have been soundly criticized for allowing the amendment to pass.

Further legislative development is likely to be under the impact of the country's Association Agreement with the EU. Bulgaria has already started a dialogue with the EU on harmonization of environmental legislation, standards and regulations. Together with the other accession countries, Bulgaria participates in ministerial level consultations with the European Commission. Within the country, the MOE is represented in the Commission for European Integration, which reports directly to the Ministerial Council and is very active in formulating the integration strategy of the country. Both timing and costs of compliance with the EU standards and regulations, including those on the environment, are expected to be subject to a serious assessment in the Commission.

Implementing organizations

Environmental management is influenced indirectly by the activities of several different ministries and committees in the Bulgarian Government. The Ministry of Environment, formed in 1991 from the former Committee on Nature Protection, is the primary administrative body charged with developing and carrying out state environmental policy. Under the Environmental Protection Act, the MOE defines Bulgaria's national environmental policy, monitors and enforces regulations related to protecting the environment, co-ordinates the activities of other ministries as they impact the environment, establishes fees for the use of natural resources in consultation with relevant ministries, designates and manages protected areas, and oversees international environmental commitments and protocols.

The MOE acts primarily through a two-layer structure comprised of the central and regional offices with the Regional Environmental Inspectorates (REIs). A third research component is housed in the National Centre for Environment and Sustainable Development (Figure 2.5). Total staff numbers are 780 – 134 in the central ministry offices, 516 in the regional inspectorates, and 130 in the National Centre for the Environment. Staff are experienced and highly qualified (70% have university degrees). Since 1991, the MOE has significantly strengthened its capacity by creating new departments, increasing the number of staff, and providing extensive training in environmental management and policy for its staff members. The REIs, which were created during the socialist period to monitor the environment and enforce laws and regulations at the local level, also have more staff now, better monitoring equipment, and expanded opportunities for technical training funded by the extra-budgetary sources and external donors.

In 1994, the MOE established the National Nature Protection Service to oversee protected areas and co-ordinate activities impacting biodiversity and the conservation of natural ecosystems. It is a small agency, and the structure and functions of the Service are still evolving. Staff is limited, and policy planning and management responsibilities for natural systems remain fragmented. In mid-1995, the Ministerial Council reviewed the role of MOE and suggested that the Nature Protection Service be placed under the Committee on Forests, which is independent of the MOE and carries out

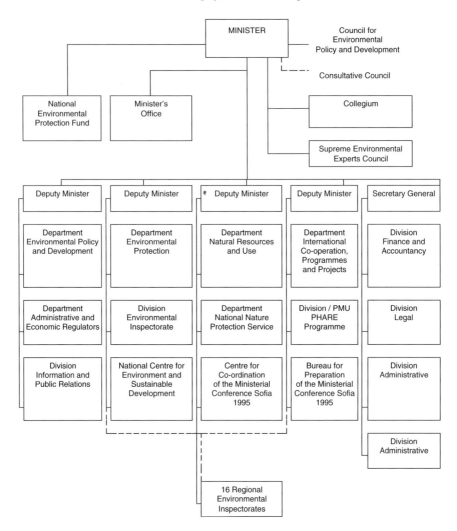

Figure 2.5   Ministry of Environment: organization chart

national forest, game and non-commercial fisheries policy. The Committee on Forests already has trained personnel in its local branches that could staff the Service. The suggestion to combine the two organizations was opposed by Bulgarian experts and by the donor community supporting biodiversity protection in Bulgaria. They were concerned with the potential for a conflict of interest between the goals of the Forestry Committee and the Nature Protection Service. After an open debate, the parties involved have tentatively agreed that the MOE will retain managerial responsibilities over the Service. This debate, however, illustrates a broad ongoing problem – how to streamline the environmental responsibilities of key government agencies. The MOE was restructured in 1995 and again in 1996 to provide better internal and external co-ordination among agencies. It is too early to judge whether the changes will be effective.

In the short and medium term, the key management issues facing the new Nature Protection Service and the MOE are related to inter-agency co-ordination, public involvement, and financing. Mechanisms need to be developed to harmonize the activities of other government agencies that directly and indirectly impact natural resource management with the overall environmental goals of the Government. Public education and conservation information need to be promoted and better funded. Finally, a sustainable internal means of financing conservation programmes and protected area maintenance needs to be found. Most of the funding for nature conservation currently comes from abroad: American, French, Swiss, and United Kingdom donors and the Global Environmental Facility. Future sources of revenues could include user permits and fees, sustainable resource extraction, and leverage of external funds through programmes such as debt-for-nature swaps.

A National System for Ecological Monitoring and Environmental Information (NSEMEI) has been established to harmonize data standards with EU standards, and to improve enforcement of environmental regulations. It is currently funded from the state budget and the National Environmental Protection Fund, but a means must be found to make it self-financing if it is going to survive. New monitoring equipment provided by PHARE raised the direct and indirect costs of operations, as the costs of general maintenance, chemicals, and spare parts went up significantly. Currently each autumn, monitoring programmes are cut back dramatically because funds for day-to day activities have run out. The System is now beginning to charge for administrative services, informational materials, on-demand analysis, and some labour costs. These revenues are used to support regular monitoring programmes. In addition, the Service could expand its fee basis, charging the State, municipalities, and enterprises for monitoring and analytical services.

Other state bodies directly responsible for environmental policy are the Ministerial Council and the Parliament. The Ministerial Council has the right to draft and present environment legislation to shape policy. A Parliamentary Committee on Environment exists, but was inactive in the previous Parliament because of political disputes among its members. The new Parliament, elected in 1995, has formed a more proactive committee charged with drafting and presenting environmental laws. The committee will also subject all legislation to environmental review to determine if it will have any direct or indirect environmental impact.

A number of additional ministries and government agencies are involved in environmental policy. The Ministry of Regional Development and Construction co-ordinates regional planning and development, and establishes land-use regulations that impact energy demand, industry, and the use of other resources. The Ministry of Health, particularly through the Centre for Hygiene and the Hygiene and Epidemiology Inspectorates, provides pollution monitoring and control in relation to impacts on human health. The Ministry of Agriculture oversees agricultural lands and production, administering soil conservation programmes and the restoration of contaminated soils. The Ministry of Agriculture also houses the State Fisheries Inspectorate, which monitors and manages aquatic resources. It is supported by the Bulgarian Academy of Agriculture, which oversees research and experimental stations that develop farm breeding programmes for local use and administers the country's germ plasma storage facility. The National Water Council manages water resources. The Committee on Geology and Mineral Resources controls the exploration for and extraction of mineral resources. The Committee on Tourism co-ordinates tourism development. Finally, the Bulgarian Academy of Sciences conducts environmental monitoring and research through the Institutes of Ecology, Botany, Water, Zoology, Forests, and the Centre for Hydrology and Meteorology.

In the past, there has been substantial functional overlap among government agencies in charge of environmental concerns, with little communication among them. Staff members have improved intra- and inter-agency co-ordination and collaboration in recent years. There are still problems, however. Pressed by sectoral problems, the government agencies are unable and sometimes unwilling to co-operate on environmental issues. The Inter-Ministerial Committee for Planning and Implementing Environmental Projects is an example of the difficulties encountered promoting intra-governmental environmental communications. This Committee was established as a high level body for environmental co-ordination and was chaired by the Prime Minister, but it is currently non-functional. The purpose of the Committee was to involve the Ministerial Council in designing a co-ordinated and sustainable development policy by providing a mechanism to integrate environmental policy into their considerations. Members met once under the previous government, and have not met since. Committee activities were perhaps not well-thought through. The Committee was charged mainly with prioritizing environmental investment projects for submission to external donors, rather than as a policy development group. The Committee duties should either be reformulated from project selection to policy orientation and reshaped into an environmentally sustainable development committee, or the Committee should be abolished.

2.3 FINANCING ENVIRONMENTAL IMPROVEMENTS

Trends in environmental expenditures

Shrinking industrial production has meant decreased pollution, but it has also meant that less money has been allocated for environmental programmes by each succeeding government. Since 1993 the share of environmental expenditure in the GDP has almost recovered its pre-1989 level, although the decrease in the real GDP meant a lower level of real environmental spending (see Figure 2.6). Competition for the limited state and donor funds available for environmental improvements has been fierce, forcing activities to be carefully targeted and prioritized. The state budget supports the central Ministry, the Regional Environmental Inspectorates, and state-financed environmental investments, such as pollution-abatement technologies for power plants and state enterprises, municipal infrastructure, mass transit programmes, and municipal wastewater treatment plants. In a major step forward, MOE is negotiating reforms in the 1996 budget in the state programme of environmental subsidies. In the past, these subsidies have been distributed indiscriminately among all municipalities. In the future, MOE wants these subsidies focused on a limited number of high priority investments in which the environmental impact will be particularly significant. For example, subsidies should go to environmental infrastructure projects that will benefit large numbers of people, or towards finishing key projects that are near completion (94.4 billion leva (US$1.3 billion) are currently bound up in incomplete capital projects for wastewater treatment plants).

Environmental regulations are being better enforced, due to better staffing in the environmental agencies, more consistent monitoring and enforcement of standards, and increased effectiveness of the general environmental policy regime. As a result, the big point source industrial polluters are increasing their resources devoted to environmental controls in order to avoid closure or censure (see Figure 2.7). State subsidies for environment are decreasing, while the share of enterprise and municipal resources put into environmental controls is growing. For example, the non-ferrous metals enterprise near Plovdiv, which was temporarily shut down on environmental grounds, now spends 160–180 million leva (roughly US$2.5 million) annually on pollution reduction and monitoring – a significant investment. The Plama-Pleven Oil Refinery now spends about 80 million leva (or US$1.1 million) annually to clean up its waste effluent into the Vit River.

Environmental funds

As in other former socialist countries, extra-budgetary funds (EBFs) are well established in Bulgaria. These funds are financial accounts held outside of the framework of the state or municipal budgets, funded from assigned revenues (taxes or other sources, e.g. fines and charges), and earmarked for particular programmes or activities. Currently 14 such funds exist, some of which were inherited from the centrally planned system; others were created to facilitate the transition to a market economy or to cushion the social impact of restrictive budgetary policy. Although justified in some circumstances, the EBFs usually distort public financial management. In the case of environmental protection,

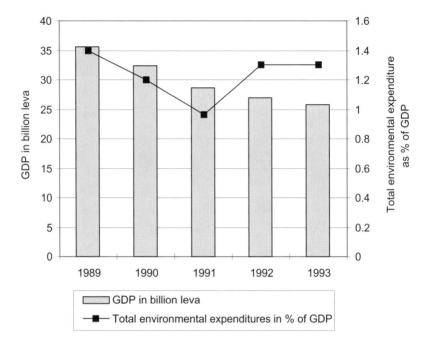

Figure 2.6   Environmental spending decrease in real terms: trends in real GDP and environmental expenditure (as % of GDP) in Bulgaria, 1989–93. *Sources*: Ministry of Environment (1995); *Statistical Yearbook of Bulgaria* (1995)

however, two factors warrant existence of the EBFs: (i) the lower priority assigned to environmental protection during the transition to a market economy; and, consequently, (ii) the decline in environmental spending and share of environmental expenditure in the state budget. The lax application of environmental regulations and failure to recover full costs for environmental services (such as sewerage, wastewater treatment and waste disposal) create the need for subsidies or soft loans to encourage pollution prevention and finance critical environmental investments.

According to the Environmental Protection Act of 1991, earmarked revenues from pollution fines and charges are directed to the National and Municipal Environmental Protection Funds: 70% of the revenues from fines go to the National Funds and 30% to the Fund in the violator's municipality. The charges and taxes currently imposed or proposed include water pollution charges, product charges (for leaded gasoline), tariffs for visiting protected areas, and fees for the collection of medicinal plants. Since their establishment, Fund income has risen annually. As indicated above, sources of revenues for the National Fund include import taxes, permits, fines, interest on loans

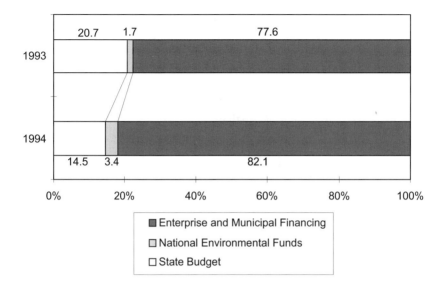

Figure 2.7 Domestic environmental expenditure in Bulgaria by funding sources, 1993–94. *Source*: *Ministry of Environment (Green Book)* (1995)

provided by the funds, loan repayment, and 5% of the state revenues from privatization.

The National Environmental Protection Fund is managed by an elected board with transparent operating procedures for the allocation of expenditures. The board is chaired by the Minister of Environment, and includes Deputy Ministers and other high level officials from relevant ministries and institutions. Supervision of environmental expenditure could be further improved through advisory committees employing broader expertise (technical, financial and managerial) and representing various interest groups, which would work in tandem with the elected board. In this way, the influence of existing institutional biases on the management of funds will be lessened. In addition, the national and municipal funds should be subjected to yearly review by independent financial auditors.

National Environmental Fund disbursements are directed towards environmental monitoring, grants and interest-free and soft loans for municipal and industrial environmental investments. They also subsidize research and NGO activities such as community education. There is public pressure to shift resources from the national to the municipal funds and increase the share of decentralized revenues for environmental protection. Unfortunately, the administrators of the municipal funds have not been able to ensure that this money is spent exclusively on environmental projects. Environmental funds are regarded as fungible by municipal authorities, and money is often transferred into non-environmental budget lines, or spent on activities misrepresented as environmental. Larger municipalities appear to be better at properly managing these funds than smaller municipalities. To a large extent, the difficulties in decentralizing environmental financing reflect the overall constraints and slow progress of decentralization in Bulgaria.

The system of pollution fines and fees that support the Funds still does not reflect the costs of pollution abatement, much less the social costs of environmental degradation. The level of fines has actually declined in real terms since 1989 as a result of inflation, and the bad financial condition of the worst polluters has made enforcement of the fines difficult. About half of the fines levied in the last two years have been actually collected. Consumer charges, such as the secondhand car tax and privatization revenues, appear to be more likely to be collected. Table 2.7 lists some of the fees and fines used to control environmental behaviour.

Trends in the allocation of resources

In the past, MOE and its predecessor were biased towards water quality protection because they were responsible for building wastewater treatment plants. The Ministry actively pressured for state subsidies for water-quality protection and directed its extra-budgetary funds to that sector. In the early 1990s, as MOE directed more energy to staff training and policy planning, the use of resources for environmental improvements became better related to the environmental problem priorities. Since 1992, the share of budget outlays for water projects has decreased, and more money has been spent on air protection and waste management. The water sector, however, still receives the largest share of environmental expenditure (44.2%). As farms have begun to be privatized and large-scale agriculture has cut back on agricultural inputs, the share of soil protection expenditure has also decreased (Figure 2.8).

It should be stressed that less than half (40%) of the expenditure in environmental protection is directed to new investments, and the rest is used for maintenance and operational costs, as well as monitoring and control. Being the biggest polluter, the industrial sector is also the biggest contributor to environmental spending (66.3% of all costs).

Foreign assistance

Because of its serious environmental problems and active environmental policy, Bulgaria receives relatively high external assistance. The EU PHARE Programme has been the primary source of external financing for environmental programmes. From 1990 to 1994, 29.7 million ECUs were committed for environmental protection with a particular focus on purchasing up-to-date monitoring equipment. Of these committed funds, 18.9 million ECUs have

Table 2.7   Environmental protection fees and fines in Bulgaria

Charge or fee	Collecting authority	Targeted group
Solid waste fees	Municipal authorities	Households and enterprises
Wastewater discharge	Local water supply and wastewater treatment companies	Industrial and domestic users
Water use fees	Local water supply and wastewaster treatment companies	Industrial and household users
Liquid fuels tax	National finance authorities	Municipal, industrial and household users
Pollution charges and fines	National and municipal authorities	Municipal, industrial and individual polluters
Automobile import duties	National finance authorities	Consumers purchasing cars more than ten years old
Differentiated gasoline tax	National finance authorities	Consumers purchasing leaded gasoline
Administrative charges	MOE and REIs for administrative and technical services	Municipalities, industrial enterprises and households

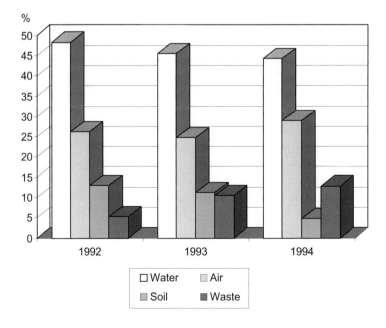

Figure 2.8   Shares of environmental expenditure in Bulgaria in 1992–94: main categories. *Source*: *Ministry of Environment (Green Book)* (1995)

been disbursed: 15.3 million ECUs for monitoring equipment and 3.6 million ECUs for training and technical assistance. In 1994, the critical transboundary co-operative programme between Bulgaria and neighbouring countries was funded by PHARE for equipment, environmental investments, and technical assistance. Slow and complicated bureaucratic procedures, on the part of the donor, and the limited opportunities for Bulgarians to participate in decision-making have engendered complaints by Bulgarians about PHARE's assistance programme. Recent improvements in the PHARE Programme have been appreciated, especially in light of the expected major PHARE assistance effort to the associated countries so they can meet the EU environmental require-ments over a certain period of time. This is to be negotiated with the individual country as a part of the accession process.

The biggest donors of bilateral environmental aid include US AID, US EPA, the Swiss Government, and the UK 'Know-how' fund. The World Bank and EBRD are both actively involved in lending for environmental invest-ments – the World Bank is financing completion of partially constructed wastewater treatment plants under the water sector loan, and EBRD is financing desulphurization technologies in the Maritsa power plant. The Japanese soft-loan environmental facility, OECF, has contributed US$80 million to reduce SO_2 emissions in two non-ferrous metal smelters in Asenovgrad and Elisseina. In 1995 the GEF council approved a World Bank-executed US$10.5 million grant for phasing out ozone depleting substances, the first in a series of similar projects in Central and Eastern Europe.

In 1995 Bulgaria signed a debt-for-environment swap accord with Switzer-land, thus becoming the second country in Central and Eastern Europe (after Poland) to apply this instrument. A debt-for-environment or debt-for-nature swap is a type of debt–equity swap involving the purchase of a country's debt at a discount rate on the secondary market and its redemption in return for increased environmental spending in the debtor country. Since they were first introduced in an agreement between a US NGO and Bolivia in 1987, these swaps have been used by many indebted countries to reduce their external debt while mobilizing resources for environmental protection.

The Swiss–Bulgarian debt-for-environment swap is for SFr20 million (20% of the official Bulgarian debt to Switzerland). The deal was facilitated by the World Bank, which also provided a $200 000 IDF grant to support the establishment of an earmarked Eco-Trust Fund for managing the swap funds, staff training, and developing the Fund's initial portfolio. The Bulgarian–Swiss swap accord set up spending priorities according to those adopted in the Environmental Action Programme for Central and Eastern Europe. The Fund's spending objectives also conform to national environmental priorities, focusing on pollution problems with severe impacts on public health.

There are a number of positive trends occurring in foreign assistance. The share of environmental investments in the country is rising relative to technical assistance. Donors are slowly increasing the involvement of Bulgarian experts in research, technical assistance, and project implementation. Efforts on the part of donors to support local institutions capable of providing consultancy services have noticeably energized the domestic consultancy market. For example, the Environmental Management Training Centre, sponsored by the US Government, established a group of trainers able to develop the Bulgarian capacity for providing environmental services. Through them, the Centre has significantly contributed to the development of environmental management skills in the country. Still more needs to be done, however, to improve the dialogue between external donors and internal experts, to involve environmental NGOs, and to support the development of Bulgarian environmental consultants. Further steps to increase Bulgarian participation in all donor-co-ordinated programmes continues to be a matter of concern.

2.4 PUBLIC ATTITUDES TOWARDS THE ENVIRONMENT

An outlook on what people care about

The Agency for Social Analyses in Sofia assessed the public perception of environmental issues in 1994 through a representative survey. The following data regarding public attitudes are summarized from their report, *Bulgarians on Environmental Problems*. The Agency found that people regarded water pollution as their most important environmental problem, followed by air pollution, use of pesticides and chemicals in agriculture, and safety in nuclear power stations. Although the condition of the environment is little improved since the late 1980s and early 1990s, people are far less concerned with environmental problems today than they were then. They worry instead about day-to-day survival, unemployment and crime. In the early 1990s, mass protests closed down or forced production cuts in dozens of polluting enterprises, thrusting hundreds of Bulgarian workers into 'environmental unemployment'. As a result, the Government is under public pressure now to keep enterprises open and invest in clean technologies instead. Nearly 70% of the people believe that economic growth is bad for the environment (a higher percentage than in any other country in Europe), but about 82% believe that economic growth is an essential prerequisite to protecting the environment.

In the 1995 survey by the Agency for Social Analyses, one-quarter of the population identified themselves as poor. About one-third of the Bulgarians surveyed experienced shortage of funds to buy food and 40% found it difficult to buy needed medicines. Despite this widespread financial stress, 34% of the Bulgarians expressed willingness to accept cuts in their existing living standard in order to improve the quality of the environment. Over 40% of the people were willing to pay higher prices so that they and their children could live in a clean environment, and one-third would accept tax increases if the funds were spent on the environment. The survey report speculated that this significant level of public awareness was due to direct exposure of many people to environmental pollution.

Despite this willingness to sacrifice for environmental improvements, most people (over 70%) believe there is not much they as individuals can do that will significantly help the environment. The Agency for Social Analyses felt this is due to a feeling on the part of individuals that they have limited environmental knowledge, as well as their paternalistic expectations about the State – that government at all levels has the primary responsibility to correct environmental problems in the country. Only 4% of the people surveyed thought industrial management should assume some responsibility for pollution from their enterprises even though most people realize industry is among the largest sources of pollution. About 20% support the 'polluter pays' principle. People believe instead that the Government and Parliament should somehow provide the bulk of the funds needed to pay for improvements. Bulgarians may be reluctant to place the onus of clean-up on industry because of the bad financial condition of many of the enterprises on which they depend for their livelihoods. On the other hand, most Bulgarians (80%) do or would participate in environmental actions that directly affect them, such as cleaning up the areas around where they live or planting trees. People generally doubt, however, that organized protest actions are effective.

The conclusion drawn from this survey is that Bulgarians are concerned about the environment, but their concern stems mostly from specific fears about their health and exposure to toxins, rather than from fundamental ecological values or culture. Environmental literacy must be strengthened in the population. This must include shaping a sense of individual responsibility and initiative *vis-à-vis* the environment. The role of non-governmental organizations (NGOs) in this educational process is key. Bulgarian NGOs have been instrumental in sensitizing the mass media to environmental issues, training professionals in all sectors of society, and promoting community education through workshops, education courses, and publications.

Non-governmental organizations (NGOs)

There are a number of active NGOs that organize community and expert responses to environmental problems in Bulgaria. As mentioned earlier, environmental issues formed the ground for open criticism to totalitarian Socialism and triggered public resistance in the late 1980s. Many Bulgarians viewed and respected Ecoglasnost more for its stand as a non-Communist organization than for its environmental concerns. Since the early 1990s, the environmental movement has become rather fragmented with some NGOs taking a clear political path and others focusing on purely environmental or conservational issues.

In general, the previous rather negative attitude towards NGOs by the MOE and other arms of the Government is improving, and representatives from the NGO and the academic communities are being invited to participate in government activities. There is a special department in the MOE, which encourages public participation in environmental policy, contacts with the NGOs and the media. NGOs are included on the boards of the National Environmental Fund and the new Eco-Trust Fund. As a result of their efforts, NGOs are now eligible to receive funds from the National Environmental Protection Fund. Nonetheless, attitudes towards NGOs are mixed, and often government agencies tend to express preference for more co-operative NGOs. During the Ministerial conference in October 1995, the NGOs were not allowed access to plenary discussions on an equal footing with the official government delegations.

Environmental NGOs are active in their efforts to get incorporated in all major environmental activities. At the roundtable on debt-for-environment swaps conducted during the Environment for Europe conference in Sofia in October 1995, the representative for the Bulgarian NGOs stressed the importance of NGO participation during implementation of the Swap agreement. NGOs need to be included to encourage public participation in decision-making, to assure public support of the various activities under the Swap, and to facilitate transparency of the overall process. A small grants programme, funded from the Swap, is included in the final agreement and is expected to provide seed money for environmental education, research and NGO activities.

Among the NGOs, ND Ecoglasnost has been particularly active in communicating and organizing public responses to nuclear safety issues, transboundary pollution, water shortages, and the need for industrial clean-up. ND Ecoglasnost, the Environmental Management Training Centre, Borrowed Nature, and the Green Party have all participated in drafting new legislation, particularly the 1991 Environmental Protection Law and regulations for the collection, transportation, storage and neutralization of hazardous wastes. A coalition of NGOs, formed among the Green Balkans Movement, Wilderness Fund, Bulgarian Society for the Protection of Birds, Bulgarian Union for Protection of the Rodope Mountains, and Ecoglasnost-Varna, worked with the Bulgarian Academy of Sciences, the MOE, Forestry Committee and others to develop the National Biodiversity Conservation Strategy in 1993.

Bulgarian NGOs are working closely with the Regional Environmental Centre (REC) in Budapest, which also has its representation in the country. The REC is very instrumental in promoting public outreach and small-scale activities. Through the REC, both in-country collaboration and East–East know-how transfer have been encouraged.

The US-based Institute for Sustainable Communities, in partnership with Ecoglasnost, has assisted Troyan (a medium-sized city) through the process of defining and prioritizing its environmental problems, agreeing on objectives, developing action plans and implementing cost-effective strategies. This partnership is likely to be repeated in other communities in Bulgaria. In 1994, a three-year project, funded by US AID through the Harvard Institute for International Development (HIID), was established to support the development and implementation of Environmental Action Plans in two communities, Stara Zagora and Kurdzhali. The NGO, TIME (This Is My Environment) Foundation, is working with HIID on this effort.

It should be noted that many Bulgarian NGOs rely on international partnerships for conceptual and financial support. These partnerships, which usually have proven to be very useful, carry the risk of externally driven agendas. According to some Bulgarians, international NGOs have too strong an impact on their local partners. TIME is given as an example of a Bulgaria-based NGO with relatively limited local outreach. More attention to building up local constituencies *vis-à-vis* creating external partnerships is recommended. There are also concerns about the ability of local NGOs to sustain their activities in the absence of international support. The slow development of democratic institutions and of decentralization makes a case for more time (five years or more) to be spent on projects that involve these groups in order to allow the environmental institutions time to strengthen.

2.5 CONCLUSIONS

During the first half of the 1990s, both the state of the environment and the policies for environmental protection improved in Bulgaria. Trends in air, water and soil quality have been generally positive. There has been a consistent effort to establish environmental priorities, ensure a coherent and comprehensive policy framework and upgrade the environmental monitoring system. Public access to environmental information has been significantly enhanced. By and large, reform efforts in the field of environment appear to be relatively more effective than in the rest of the economy. The slow progress of economic restructuring has been the most important constraint to achieving better results in terms of environmental improvements.

Data show that changes in environmental quality have been mostly due to economic decline and will not be sustained without major efforts to ensure cost-effective measures to address high priority problems. Water quality appears to be the environmental issue of greatest public concern in Bulgaria largely because of the severe water shortages that have plagued the country in recent years. The most serious sources of water pollution in the mid-1990s are identifiable point sources such as municipal wastewater, large industrial enterprises, and small and medium-sized food processing plants. Air pollution, although generally perceived as a less serious issue by Bulgarians, has the highest negative health impacts and is likely to persist as a key area of concern in the next few years. Particulates, nitrous oxides, and sulphur dioxide are the primary pollutants emitted by both large point sources, such as power plants, smelters, and other industrial enterprises, and a rapidly growing number of non-point sources (cars in particular).

Farming is the main source of non-point water pollution, but the dramatic reduction in the use of chemical inputs and the shrinking livestock herds have reduced the seriousness of this threat for the immediate future. Land restitution and the restructuring of agricultural areas both in terms of the size and management of farms and the types of crops produced will exert new monitoring and regulatory demands on the local and national government agencies. Through the joint efforts of the Ministry of Environment, Ministry of Agriculture and environmental NGOs, the Government is making some effort to avert future problems by offering training to farmers in environmentally-friendly practices, particularly in fragile areas.

The main challenges for the environmental authorities in the short and medium term are likely to be in the areas of enforcement of regulations, incentives and compliance schedules for big point-source polluters, and improvements in controlling non-point sources. Promoting pollution prevention and sustainable patterns of industrial, agricultural and urban growth will require mainstreaming environmental concerns in the overall economic development. Although there is progress in promoting voluntary compliance, it is still a major task for Bulgaria to find a way to encourage citizen compliance and self-reporting on environmental issues – and to improve the ability and efficiency of the Government to respond to public efforts such as citizen monitoring and reporting.

The environmental policy-makers must improve communications and co-operation among agencies that form and implement macro-economic and political policies such as decentralization, privatization, prices and fiscal policy. The strategy for phasing out leaded gasoline offers a good example of cross-agency co-operation to reduce traffic-caused pollution. Political, economic, and environmental policy reforms must occur simultaneously for the changes to be robust and sustainable, but market reforms are moving slowly. Privatization is barely begun, state subsidies for polluters persist, pollution fines, although higher than they once were, are still too low, and the existing industrial structure endures. The package of economic instruments framed in the 1991 Environmental Protection Law will work only in an environment of hard budget constraints and a market economy. The increase in pollution since 1994 confirms the dependence of environmental progress on the scale and speed of economic reforms and industrial restructuring. An effective pollution policy will continue to elude Bulgaria until reliance on old ways of doing business is set aside and new economic and political standards are internalized.

Bulgaria has been relatively successful in maintaining the share of environmental expenditure in the GDP mostly because of the consistent efforts of the environmental authorities to encourage enterprise financing and to raise extra-budgetary revenues. Leveraging external assistance for environmental protection has also been effective and should be continued through pursuing innovative programmes such as the debt-for-nature swap negotiated with the

Swiss Government. The sustainability of financing conservation programmes and pollution monitoring, prevention and abatement programmes must be guaranteed through the evolving system of user or polluter permits, charges and fees.

Bulgaria should actively pursue environmental co-operation with the EU and within the group of associated countries. Harmonization of standards and regulations with those of the EU and development and implementation of harmonization schedules will be one of the most important factors for environmental improvements in the short and medium term. Bulgaria, as with the other accession countries, should design and implement economy-wide efforts for least-cost harmonization. It will be a major challenge both to complete the current package of environmental legislation needed to implement the Environmental Protection Law, and to undertake harmonization with the EU legislation (especially since parts of the latter are also subject to revisions).

Decentralization of environmental services is as critical to environmental management as it is to other sectors of the society. The municipal authorities, as well as the REIs, would like to have more independence in performing their functions and in assuring that funding is less dependent on decisions of the central authorities. More co-operation between the REIs and municipal authorities may help local officials gain a better understanding of the need to encourage environmentally-friendly behaviour in both small and large enterprises. For this purpose, the REIs should be strengthened. One of the potential steps in this direction might be to increase the number of REIs (for example, by extending them to all former oblast centres) to improve their ability to communicate with all municipalities on their territory. Closer links would also make it easier for the MOE to develop incentives targeted to individual locales and firms. Until fiscal decentralization is accomplished, however, locally responsive incentive systems will be impossible to implement.

Committed scientists, government administrators, academics and activists are continuing to pursue environmental policy reform, although a sense of environmental urgency has slipped from the public consciousness. Lack of co-ordination among interacting government agencies, insufficient public involvement in decision-making, difficulties with monitoring and enforcing conservation regulations, and severe budgetary restrictions continue to hinder effective management. To overcome these constraints, public education must be better funded, including programmes that teach individual responsibility for guarding environment, and that support enterprise managers' and workers' initiatives to reduce pollution. To generate a constituency for ongoing environmental development, however, environmental literacy needs to be enhanced through the media, schools, and activities of advocacy organizations.

We cannot emphasize enough that a co-ordinated solution must be found to both the economic and environmental problems. Economic restructuring has to occur to move the environmental reforms forward. As recovery of the Bulgarian economy is a prerequisite to solving the country's environmental problems, cleaner industrial growth should be encouraged through government incentives and investment programmes. To ensure sustainable improvements, environmental management must be integrated into the social and economic transition currently underway. While much has been accomplished, much more still needs to be done, both to accelerate the overall reforms and to incorporate environmental concerns into the economic transition. Clear commitment to reforms and political will are essential, and the right time to apply them is now.

REFERENCES AND FURTHER READING

Agency for Social Analyses (1995) *Bulgarians on Environmental Problems*. Prepared for the Third Ministerial Conference, 'Environment for Europe'. Sofia, 23–25 October 1995.

Biodiversity Support Program (1994) *Conserving Biological Diversity in Bulgaria: The National Biological Diversity Conservation Strategy*. Biodiversity Support Program c/o World Wildlife Fund, Washington DC.

Commission of the European Communities (1995) *Transboundary Pollution Romania–Bulgaria*. PHARE Final Report, CEC.

Friends of the Earth (1995) *EAP Policy Paper from European NGOs*. Prepared by Milieukontakt Oost-Europa and Friends of the Earth Europe for the Third Ministerial Conference, 'Environment for Europe'. Sofia, 23–25 October 1995.

Lovei, M. and Levy, B.S. (eds) (1995) *Lead Exposure and Health in Central and Eastern Europe – The Impact on Children: Evidence from Hungary, Poland and Bulgaria*. Implementing the Environmental Action Programme for Central and Eastern Europe. The World Bank, Washington DC.

MOE (1995) *Annual Report on the State of the Environment in 1994 (Green Book)* (in Bulgarian). Ministry of Environment, Sofia.

National Statistical Institute (1995a) *Environment '94* (in Bulgarian). NSI, Sofia.

National Statistical Institute (1995b) *Socio-Economic Development of Republic of Bulgaria in '90–'94* (in Bulgarian). NSI, Sofia.

National Statistical Institute (1995c) *Statistical Yearbook of Bulgaria* (in Bulgarian). NSI, Sofia.

Regional Environmental Centre (1995) *Status of National Environmental Action Programs in Central and Eastern Europe*. Case Studies of Albania, Bulgaria, Croatia, Czech Republic, Hungary, Latvia, Lithuania, FYR Macedonia, Poland, Romania, Slovak Republic, and Slovenia. Budapest, The Regional Environmental Centre for Central and Eastern Europe.

Task Force for the Environmental Programme for the Danube River Basin (1992) *Strategic Action Plan for the Danube River Basin: 1995–2005*.

UNECE–FAO (1994) *Forest and Forest Products Country Profile*. Geneva Timber and Forest Study Papers, No. 1. United Nations Economic Commission for Europe and the Food and Agriculture Organization, New York.

UNEP, UNDP and World Bank (1993) *Saving the Black Sea*. Programme for the Environmental Management and Protection of the Black Sea. UNEP, UNDP, and World Bank, Washington DC.

Vari, A. and Tamas, P. (eds) (1993) *Environment and Democratic Transition: Policy and Politics in Central and Eastern Europe.* Kluwer Academic Publishers, Dordrecht, The Netherlands.

World Bank (1992) *Bulgaria Environmental Strategy Study.* Report No. 10142-BUL, 17 March. The World Bank, Washington DC.

World Bank (1994) *Bulgaria Environmental Strategy Study Update and Follow-Up.* Report No. 13493-BUL, 30 December. The World Bank, Washington DC.

CHAPTER 3

Czech Republic

Bedřich Moldan

3.1 BEFORE AND AFTER NOVEMBER 1989

The Communist ideas stemmed from eighteenth- and nineteenth-century visions of a socially just *and* equitable society. However, from the very beginning it was clear that such a society could never be realized in a democratic way. The Communist ideas were never shared by more than a tiny minority of elitist intellectuals who regarded themselves as saviours of the world. As the world proved reluctant to be saved, a brute force was applied. Eventually the elitists' saviours succeeded and imposed their rule on to others. The people were forced to follow – at any cost.

And the cost was indeed high. The vision of an ideal society quickly proved unrealistic. The established power structures were used to implement other goals. The main one was dominance over all people accompanied by coercion and imperialism. The Communist empire, especially under Stalin and his successors, evolved into what President Reagan rightly called 'The Evil Empire'.

Totalitarian rule was oppressive and arrogant towards people and nature. Neither was considered to have any value. Nature was a worthless entity, an enemy to be conquered, a sheer source of minerals and other resources. Natural resources acquired value only through human labour. If such a value were not added, the original value of nature was considered to be zero. An almost absolute misunderstanding of nature, natural resources and environment is a logical consequence of such a position. The complete neglect of the human environment is only the final result of such reasoning.

Czechoslovakia became a part of the Communist empire through a coup in February 1948. This date marked the beginning of the most disastrous period of Czech modern history.

During the 1930s, Czechoslovakia was one of the most advanced countries of the world. The damages during the Second World War were only minor

The Environmental Challenge for Central European Economies in Transition.
Edited by J. Klarer and B. Moldan. © 1997 John Wiley & Sons Ltd.

ones, and so the country had all the prerequisites for a rapid and successful post-war development. Unfortunately, the Communist coup in 1948 started a deep decline of the country which lasted for 41 years. The deterioration of Czechoslovakia was universal. Democratic structures were thoroughly annihilated. The economy was nationalized to the last shopkeeper and farmer with disastrous consequences. The educational system was badly damaged. Included in the long list of victims of Communist rule has been the environment.

The decline of Czechoslovakia was similar to that of other countries of Central and Eastern Europe under Soviet domination. Compared to the other countries, the Czechoslovak decline was indeed steeper. The overall development level at the beginning of the Communist period was substantially higher here, but in the end the level of all the CEE countries was generally very low and, at the same time, also rather similar.

Specific to Czechoslovakia was its long and excellent tradition in the heavy machinery and steel industry. Stalin was known for his obsession for steel and arms, and he was quick to exploit the Czech tradition recklessly and thoroughly. He forced Czech officials to expand Czechoslovakia's capacity to the limits. Czechoslovakia became the blacksmith of the Soviet empire. It ended up with almost a world record of steel production in per capita terms, enormous tank and other weapons factories, and so on.

One of the most devastating activities was the uranium industry. Uranium deposits are numerous in the Czech Republic and several of them were very rich. One of them – Jáchymov (Joachimstal) – has been known since the Middle Ages and was made famous by the seminal work of Madame Curie. Her radium originated here. After the Communist coup in 1948, the Soviets were very quick to start a massive exploitation of the Czech uranium deposits. Uranium ore was mined in enormous quantities, the ore concentrate produced on the spot and shipped to the Soviet Union. The first Russian atomic bombs were made of Czech uranium. The mines and processing plants were infamous sites of prison labour where mostly political prisoners suffered heavily. There are no available records of volumes of uranium mined and shipped to the USSR. Everything was controlled by the Soviets and data were strictly top secret.

Uranium has been mined in the Czech Republic both by conventional mining methods and by chemical leaching directly in the underground deposits. Conventional mining methods lead to devastation of the landscape by mining waste tips and to deterioration of water quality in surface watercourses. The environment is also endangered by settling ponds in the vicinity of uranium processing plants. These ponds contain several tens of millions of tonnes of sludge with a high concentration of residual uranium, $226Ra$, and a number of other detrimental elements.

The overall result of the Communist rule during more than 40 years was a badly deteriorated environment and over-exploited natural resources.

Czechoslovakia was one of the countries with most intense exploitation of mineral riches. Each square metre of the territory yielded about 6 kg of materials that were subsequently transformed in waste, partly toxic. The production of solid waste corresponded to about 35 tonnes per capita per year.

The most disturbed component of the environment was the air. The most prominent pollutants were sulphur dioxide, nitrogen oxides and particulates. According to the best known of them, sulphur dioxide, Czechoslovakia occupied the second place in Europe according to the emissions per capita or per inhabitant. (The first was the German Democratic Republic.) The average yearly concentration in the air was between 15 μg m^{-3} at the cleanest locations to about 100 μg m^{-3} in the most polluted areas. In the latter, the relative frequency of exceeding the highest daily allowance of 150 μg m^{-3} (a hygiene standard) was more than 30%. In some areas, frequent episodes of winter smog were occurring. During these periods, extremely high concentrations of pollutants were recorded. For example, during a particularly severe episode in January 1982, the maximum daily average of sulphur dioxide in Prague reached 3193 μg m^{-3} and in the town of Osek in north-west Bohemia 2440 μg m^{-3}. Czechoslovakia was one of the larger exporters of sulphur dioxide emission into neighbouring and also distant European countries.

Emissions and corresponding ambient air concentrations of sulphur dioxide are not only important *per se*. As they correlate with other air pollutants having the same source – combustion processes – with other harmful substances such as toxic organic compounds and heavy metals, they also indicate the level of air pollution in general. The main source of environmental pollution in the Czech Republic was heavy industry and power production which is concentrated within a relatively small number of areas. These territories gradually became large 'hot spots' of environmental pollution. The largest and most important are the coal basins of north-west Bohemia (lignite) and north Moravia (black coal). In addition to this, a very polluted area is central Bohemia around the capital of Prague and the city itself.

Pollution was of course not limited to these areas. Virtually every larger city or town had polluted air. Agriculture everywhere creates severe water pollution and other problems. Many towns and villages had no adequate waste-water treatment plant. Nevertheless, owing to the high concentration of polluting activities within the most affected industrial areas, the differences in the quality of environment among different parts of the Czech Republic were substantial. In the Czech Republic, rather unpolluted pristine areas still existed that were influenced essentially only by background air pollution common in most of Europe, e.g. in south Bohemia (Šumava Mountains).

The state of water quality was not better. Only 17% of watercourses were classified as clean (Class I) while 34% were within Class III and 22% within Class IV (extremely polluted).

Other components of the environment suffered in a similar manner.

In November 1989, the 'Velvet Revolution' in Czechoslovakia overthrew the Communist system. The former Czechoslovakia began a deep transition process towards democracy and the free market economy. This process proved to be a difficult and painful one, accompanied by a sharp drop in economic output which created hardships for large segments of the population. The gross domestic product decreased by about 15% in 1991 followed by a further decline of 7% in 1992.

The year 1993 was marked by the division of Czechoslovakia into the Czech Republic and the Slovak Republic. This brought an additional negative economic impact. In contrast, the GDP only dropped by 0.3%. The decrease of industrial production was even greater. The volume in 1993 was only 63% of the 1990 level. The year 1994 was the first one with modest economic growth. In 1995, the GDP increased by 4.5%.

The sharp decrease in industrial production was not only inevitable but mostly even beneficial. It was partly caused by a sharp reduction of heavy industry. Because the industrial sector is the main culprit of environmental pollution in the Czech Republic, the impact of its reduction on the environmental situation was markedly positive. Surprisingly enough, it was not accompanied by high social tensions. A relatively low level of unemployment was documented. In the most affected regions this reached 20% for a limited period of time. The national average did not exceed 5%. At the beginning of 1996, the unemployment rate was less than 3%.

After six years of the transition, at the beginning of 1996, the essential steps of the transformation already had been taken. The second democratically elected government was finishing its term. The legislative framework of a civic, free-market society was basically completed. A large majority of property, including big industrial plants, was privatized. The process of restitution gave property which had been socialized by the Communists back to people.

The transformation and restructuring of the economy has been an orderly process governed by laws. From the very beginning of the transition period the environmental rules, limits and standards were never softened, just the opposite. All newer laws were stricter than the older ones and even the old ones have been more fully enforced, e.g. at the end of the Communist period there existed about 2000 so-called 'exceptions' from the Water Law. That meant the polluters – all of them state-owned enterprises – were exempt from the liabilities. All these exceptions were cancelled during the first half-year of the new regime.

The economic transformation brought about a decrease of mining of lignite and coal and other minerals. The structure of primary energy sources has shifted from lignite and coal towards gas. The production of steel was almost halved. The agricultural production dropped both in absolute terms and relative to GDP. Most of the changes have environmentally-beneficial effects.

The relatively slow but steady progress during the second phase usually does not attract public attention. Nevertheless the improvement of the quality

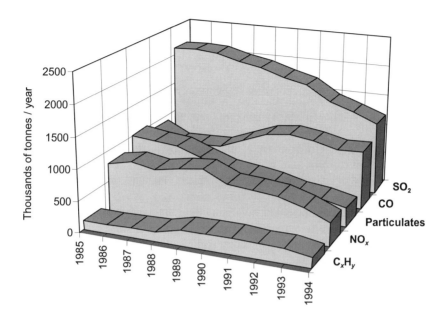

Figure 3.1   Emissions of principal air pollutants in the Czech Republic, 1985–94.
Source: Ministry of Environment

of air, water, urban spaces, living nature and other components of the environment is visible. On the other hand, the environmental deterioration during the Communist decades was so deep that the overall situation is still bad.

3.2 ENVIRONMENTAL POLLUTION AND ITS CAUSES

The pollution of air is still considered to be the number one environmental problem of the Czech Republic. It culminated in the late 1980s: 2.4 million tonnes of sulphur dioxide were emitted in 1982. Since then, the amount of SO_2 emitted has gradually decreased. The greatest decrease has occurred since 1990, primarily as a consequence of changes in the economy.

Between 1989 and 1994, total emissions of sulphur dioxide fell by 36%, and dust emissions fell by 49%. Emissions of hydrocarbons dropped by 37% and emissions of carbon monoxide decreased slightly. The emissions of carbon dioxide decreased by 15% between 1990 and 1994. The overall situation is depicted in Figure 3.1.

Ambient air concentrations of sulphur dioxide and particulates decreased especially in the most affected districts of north-west Bohemian region. Compared to yearly average SO_2 emissions of over 100 $\mu g\ m^{-3}$ in the 1980s, the corresponding value for 1994 was about 30 $\mu g\ m^{-3}$.

However, no decrease in ambient concentrations of nitrogen oxides was observed. In contrast, in some regions (e.g. Prague) the concentrations are even slightly higher now due to the rapid increase of the number of cars. The yearly average emissions of NO_x are about 60 μg m^{-3}.

Surface water pollution also decreased significantly between 1989 and 1994. During this five-year period, the total pollution discharges from point sources were reduced by the following percentages: BOD5 by 55%, insoluble substances by 43%, crude oil substances by 64%, inorganic salts by 25%, acidity/alkalinity by 72%. Unfortunately, the significant decrease in the use of manufactured fertilizers did not produce a corresponding decrease in the concentrations of nitrates and phosphates (which contribute to the eutrophication of some water flows and reservoirs). Water flows categorized as highly and severely polluted still constitute 34% of the total water flow length. Since 1990, certain regions have experienced a slight deterioration in the quality of their drinking water. In some cases, tests indicated that permitted limits were exceeded for groundwater from shallow and deep wells (in particular nitrates, crude oil products and CODs). Over this time period, water consumption and the amount of discharged wastewater has decreased considerably. Currently, 84.7% of the population receive their potable water from public water mains and 73% are connected to public sewage systems. As of 1994, 84.5% of the wastewater in public sewage systems was at least partially treated before being discharged.

The landscape is negatively affected by environmentally adverse development practices by agriculture and by many sites of potentially dangerous waste dumps. The total amount of 187.0 million tonnes of waste produced in 1991 in the Czech Republic included 48.3 million tonnes of agricultural waste, 44.6 million tonnes of industrial waste, 17.0 million tonnes of power production waste (not including radioactive waste), 4.2 million tonnes of communal waste and 61.7 million tonnes of other waste. Prior to 1991, there was no legal framework for waste management. Since then several legal norms were adopted, and the situation is improving rapidly. There are many new incinerators or other facilities under construction or already in operation as well as many recycling schemes. However, about 7000 waste dumps, including those no longer in use, are a potential danger to the environment. The total number of locations in the Czech Republic where there is a potential danger of soil and water contamination has been estimated at between 10 000 and 15 000. They include industrial sites, waste dumps and many other sites. Prominent among them are the areas of former Soviet Army bases. Environmental detriment caused by the occupation of the Czech Republic by the Soviet Army at 74 locations has been estimated to correspond to 2370 million CK (US$ 91 million).

The soils of the Czech Republic are mostly fertile, productive ones. There is no serious contamination by heavy metals or other harmful substances. On the other hand, the biological and physical status of the soils is not excellent.

Table 3.1 Protected areas in the Czech Republic (as of 1.1.1996)

Category	Total number	Area (hectares)
National parks	3	111 920
Protected landscape areas	24	1 042 365
National nature reserves	118	26 441
National nature monuments	100	4801
Nature reserves	542	22 675
Nature monuments	958	27 575

Source: Agency for Nature and Landscape Protection in Czech Republic.

Excessive use of fertilizers and pesticides in the past have resulted in the decline of soil biological activity. Heavy machinery and environmentally-unfriendly cultivation practices have caused negative changes in physical structure. About 40% of all arable land is affected by erosion (mostly water erosion) to a variable degree and requires protection.

Biodiversity has been markedly reduced due to industrial pollution and unfavourable agricultural practices with heavy use of chemicals, too big fields and too large a percentage of arable land.

The decreased intensity of agricultural production and decreasing trends in the use of industrial fertilizers and plant-protection chemicals within the transition period are leading to a gradual improvement in the biological diversity of ecological systems. Many invertebrates (butterflies, beetles) are reappearing in a great many localities along with the return of insect predator, bird species. The renewal of the natural association of species is also assisted by the establishment of new hedgerows, bush game refuges, scattered green areas that divide ploughed fields and limit the extent of erosion on agricultural land. These changes have favourable consequences for the numbers of game animals, especially small furry and feathered animals in field areas.

The pivotal element of the Czech nature protection strategy is a comprehensive system of protected areas. These have different legal status. The highest degree of protection is assigned to nature reserves, the lowest degree is for protected landscape areas. An overview is provided by Table 3.1.

Czech forests cover 33% of the country's surface and are subject to heavy environmental damage due to extensive air pollution, namely acid rain and acid deposition. Air pollution has affected more than 60% of forest stands covering virtually the whole northern half of the Czech Republic and all mountainous areas. Table 3.2 gives an overview of forest damage. Considerable damage to forests is caused by large game animals (mostly red deer). About 5% of the entire forest area is damaged by ungulates. This is also one of the reasons why an increase in deciduous trees is prevented since the animals have a food preference for them.

The pollution and other environmental damage are associated with economic activities. The most crucial economic sector is coal mining and energy.

Table 3.2　Forest damage in the Czech Republic by air pollution

Degree of damage	Damaged conifers % of total area			Damaged deciduous % of total area		
	1985	1990	1993	1985	1990	1993
No damage	50.2	38.6	36.6	89.4	54.6	42.9
First symptoms	28.6	35.0	37.0	10.6	36.8	46.7
Slight	13.1	19.0	20.0	–	8.0	9.9
Medium	6.1	5.6	5.1	–	0.6	0.5
Strong	1.4	1.3	1.0	–	–	–
Very strong	0.3	0.2	0.2	–	–	–
Dying	0.2	0.1	0.1	–	–	–
Dead	0.1	0.1	–	–	–	–

Source: Forest Management Institute, Brandýs and Labem.

Table 3.3　Structure of primary energy sources in the Czech Republic in 1992

Source	Petajoules	Per cent
Coal and lignite	1120	59
Oil	320	17
Natural gas	280	15
Nuclear and water	170	9
Total	1890	100

The energy supply in the Czech Republic is based on domestic solid fossil fuels which are mostly lignite and, to a lesser extent, hard coal. The structure of primary energy sources is given in Table 3.3.

Lignite and hard coal represent the most important commodity from the point of view of economic and environmental impacts. In 1994, the total amount of lignite mined in the Czech Republic was 67.8 million tonnes. For hard coal it was 18.7 million tonnes. These amounts were significantly lower compared to 1988: lignite was 71% and hard coal 68% of the 1988 figures.

Most of this amount was consumed domestically. In 1994, 5.3 million tonnes of lignite and 6.5 million tonnes of hard coal were exported. On the other hand, 1.7 million tonnes of hard coal were imported.

A typical feature of the economy of the Czech Republic common to all former socialist countries is a relatively high energy intensity. Measured by energy necessary for the production of a financial unit of GDP, the Czech Republic uses about two to three times more energy than comparable European countries (Germany, The Netherlands, Sweden) and up to five times more compared to the most 'energetically' advanced ones (Japan, France). Unfortunately during the early years of transition, this already unfavourable situation was even worsened. One only can speculate regarding the probable reasons. Almost certainly the main one is the relatively low energy prices that are not compensated by any administrative or economic instruments for stimulating higher energy efficiency.

Lignite mining is mostly localized in north Bohemian and Sokolov in open-pit mines. This type of mining was originally concentrated in the outcrop parts of the coal seams. In the 1960s, extensive boundary and central parts of the coal deposit area were uncovered and excavated to a depth of 150–180 metres. This open-pit mining destroyed the original environment as a whole, including accumulations of groundwater and the landscape. A total volume of approximately 185 million m^3 (about 400 million tonnes) of overburden rock had to be removed in the mining of 66.9 million tonnes of brown coal in the two areas in 1993. This overburden is deposited on enormous tips mostly outside the open-pit mines. The overall area taken over for mining and tips in the north Bohemian coal-mining area equals 26 000 hectares of which the area of the internal and external tips equals approximately 8000 hectares. The mining has left residual pits with a total volume of 3.5 million m^3.

The hard coal mining area is located in north Moravia around Ostrava and Karviná. Underground mining was carried on for more than 100 years at several levels (beds), each above the other without the introduction of effective measures to limit the surface impact of undermining. Subsidence of the surface of the terrain has influenced an area of 14 300 hectares from 1983 to 1992. This area will probably increase to 17 900 hectares by the year 2050.

Lignite and hard coal are mostly used as fuel. To a lesser extent, hard coal also serves as a raw material for production of coke that is used not only as a source of energy but also for production of steel. While the hard coal mined in the Ostrava and Karviná regions is a relatively good energy raw material, the north Bohemian lignite is not. It is a very inefficient and polluting source of energy. From the environmental point of view, the natural endowment of this relatively abundant source is viewed as a disaster. Even if the quality of lignite varies within relatively wide margins, it has unfavourable common characteristics: low content of energy, high content of ash and water and high content of sulphur. The sulphur concentration varies from slightly above 1% to 10%, sometimes even more. The average concentration is about 2.5%. The burning of lignite causes about 85% of sulphur dioxide emissions in the Czech Republic. It is by far the most important source of environmental pollution in the country. Overall, the production of 1 PJ (petajoule) of thermal energy in the Czech Republic is accompanied by the following emissions: 400 tonnes of particulates, 1800 tonnes of sulphur dioxide, 550 tonnes of nitrogen oxides and 115 000 tonnes of carbon dioxide. All these numbers are among the highest in the world.

Among other industrial sectors, iron metallurgy is very polluting. Its impact is both direct and indirect. The direct impact of steel production, for instance, causes air and water pollution. The most harmful impact is particulate emission into the air. There are also other air pollutants emitted (NO$_x$, VOC, SO$_2$, etc.). Even if the normalized emission (per kg of the product) is not exceptional, given the sheer volume of the production the total pollution by this industrial sector is great. Fortunately, pollution controls began to be installed two decades ago, and this most dangerous type of air pollution has gradually decreased. It is now essentially under control by means of the existing air pollution law that combines charges and administrative measures (standards). The production processes are steadily (albeit rather slowly) being modernized and equipped with more or less adequate abatement technologies.

Another environmental impact is connected with the generation of solid wastes. They are categorized within the industrial waste category. Part of these are hazardous because of the high concentration of heavy metals. To indirect impacts also must be added those connected with energy production, coal mining and coke production. All of them are substantial. Coke production is an enormous source of air pollution as very obsolete and polluting methods still are partly used in Ostrava. Air pollution from this source is particularly bad in major industrial centres and their surroundings. For instance, the air pollution from Ostrava destroyed many of the forests in the nearby Beskydy Mountains.

The chemical industry is comprised of several large petrochemical plants and numerous middle-sized and small firms producing many kinds of products. The main environmental impact is caused by emissions to the air (e.g. volatile hydrocarbons) and discharges to the surface waters. Many very large factories do not yet have adequate wastewater treatment plants, but virtually everywhere the lacking ones are under construction. In the last two to three years, several large water treatment plants were finished with a positive impact on the water quality especially that of the River Elbe.

The pulp and paper industry is still one of the prominent water polluters. In 1990, the two most important sources were Sepap Štětí and Biocel Paskov which produced 3400 tonnes of BOD5 (about 2% of the total BOD5 load of the entire Czech Republic). However, in recent years a significant improvement was achieved. For instance, the instalment in 1992 of an efficient wastewater treatment plant in Větřní (a large paper mill near Český Krumlov, south Bohemia) has had a positive impact on water quality as far as Prague (about 200 km downstream) where a decrease in chemical oxygen demand was observed.

The food processing industry has an impact primarily on water quality. It also produces hazardous biological and other wastes and, on some occasions, obnoxious smells. There are numerous middle-sized and small individual units that are very diverse. Their combined adverse impact on the water quality can be considerable. It is rather difficult to control all of them effectively because regulatory rules are not developed enough and the capacity of environmental inspectors is insufficient. The economic instruments (charges for water effluents) are in place, but necessary monitoring is mostly missing.

Table 3.4   Consumption of selected fertilizers and pesticides in the Czech
Republic, 1989–93

Year	Nitrogen (kg ha^{-1} of N)	Phosphorus (kg ha^{-1} of P$_2$O$_5$)	Pesticides (total, '000 t)
1989	103	67	11
1990	86	53	9
1992	50	8	5
1993	40	13	3

The recent changes in agriculture connected with the transition period could
be considered as positive from the point of view of the environment since they
already have brought some environmental benefits. Due to complete elimination of subsidies for fertilizers and pesticides, the use of these items fell
substantially (Table 3.4). An immediate positive environmental response
already has been noticed, namely the increasing number of small game
animals and other wildlife.

The structure of the agricultural landscape is changing rapidly. After the
reduction of the size of enormously big fields which were created by socialists
in order to accommodate heavy machinery, the new landscape is more
diversified. An enhancement of ecological stability is one of the objectives of
the newly established structure. The district offices have at their disposal maps
of the 'Territorial Systems of Ecological Stability' that serve as a guide for
introduction of these changes by farmers and developers.

The harmful environmental effects of agriculture are: generation of wastes,
water and air pollution, degradation of soil, and harmful impacts on biodiversity, ecological stability of the landscape and wildlife. Agricultural wastes
are, in essence, rather important resources and were viewed as such by
traditional agriculture. Modern agriculture produces wastes that are, in
contrast, viewed as obnoxious and harmful. There are too many of them, and
they are concentrated in the wrong places. Most difficult are animal wastes,
especially pig manure. At present, the big concentrations of animals typical of
socialist agriculture have been mostly eliminated. Because of that the
problems of pig manure have diminished.

Water pollution from non-point sources, which mainly means agricultural
fields, decreased as the heavy usage of fertilizers and pesticides was reduced.
Air pollution is mostly connected with animal husbandry. Until now it has
been considered only as something unpleasant. However, animals are direct or
indirect producers of ammonia and methane, both considered as important air
pollutants. No precise information on the scope of this problem is available,
only crude estimates.

Soil degradation was a typical product of socialist agriculture. Over-
fertilization, compaction by heavy machinery, too large surface area of the

fields, over-use of pesticides, inappropriate cultivation and harvesting machines caused chemical, physical, and biological degradation of the soil. Some of the practices still persist but, in general, there is a substantial improvement during recent years.

Large fields without trees or any other greenery are conducive to reduction of biodiversity and ecological stability of the landscape. Indeed, during a decade or so before the fall of Communism, a very serious reduction of biodiversity in the agricultural landscape was observed. During recent years this process was reversed. In fact, the marked improvement in this respect is one of the most visible effects of the transition processes in agriculture.

The amount of transport decreased since 1989 mostly due to the decline of the demand for freight transportation. Transport decreased especially along railways and by domestic road transport companies. International truck transport increased and so did the average transportation distance. This led to a growth in the amount of traffic in some sections of highways and super-highways with detrimental consequences for the environment along these routes (exhalations, noise). The decrease in environmentally preferable railway transportation can also be considered unfavourable. The decrease in public transport was also partly a consequence of the closing of unprofitable routes as a result of a decrease in subsidies. The number of motor vehicles in the Czech Republic is continuously increasing but has not yet attained the figures for the European industrially advanced countries.

Between 1989 and 1994, the number of cars, trucks and buses increased by 30%. In 1994, approximately 7–8% of vehicles were equipped with catalytic converters. The growth in the use of unleaded motor fuel and the substantial reduction of lead content of the normal fuel have contributed to approximately 75% reduction of lead emission between 1989 and 1994.

3.3 ENVIRONMENT AND SOCIETY

Environmental improvement was a high priority among citizens of Czechoslovakia (the same for other Central and Eastern European countries) before and after the fall of Communism. There had been a deep concern over health and other impacts of environmental pollution. The pressure from environmentalists and concerned citizens substantially helped to break down the Communist system. After the establishment of democratic government, there was wide public support for radical environmental improvement. According to one of the very first public opinion polls conducted in January 1990, more than 80% of the Czech citizens considered the environmental clean-up the very first priority for the new government, even before economic and other problems.

Building up a functional system of environmental protection was indeed high on the priority list of the new government. The basic goal was to halt the worrisome trends of continuing deterioration of all the environmental components. The impacts on human health were the prime targets. During the years 1990–92, 12 laws were either prepared or passed through the Parliament as were other legal norms that set the basic structure and rules of environmental protection in the new democratic society. Not only control-and-command instruments were used but also economic instruments such as charges. The Fund for the Environment was established to support environmental investments. Other institutions that were established were: the Ministry of Environment with broad competencies and nine regional offices in the centre, the State Environmental Inspection, and several other institutions. Some well-established organizations like the Czech Hydro-Meteorological Institute, the Czech Geological Survey, and the Water Research Institute were incorporated into this system. Some new ones were created, such as the Czech Environmental Institute.

The expenditure for the environment gradually increased from less than 1% of GDP in 1990 (and certainly before that) to 2.1% in 1992. Most of the resources were used in environmental 'hot spots' such as north-west Bohemia, north Moravia and Prague.

Despite the initial enthusiasm and all of the efforts, no quick results emerged. At least three reasons exist for that. First, the damages were indeed great, in fact more deep than even the experts originally thought. Their causes were firmly rooted in the existing industrial and economic structure. Extensive technological 'hardware' was simply impossible to change quickly. Second, there was not enough experience and knowledge among both environmentalists and the relevant decision-makers to choose the most efficient ways of environmental improvement. Third, because of the very fact that there were no quick results, many people became frustrated and disappointed and their active concern for the environmental cause vanished. The public support was reduced surprisingly greatly and quickly. Thus, after a year or two, the first enthusiastic phase of the creation of environmental policy came to an end.

The initial environmental euphoria was neither exceptionally strong nor sufficiently lasting to compete successfully with other important tasks of the transition. People realized very quickly that even if the environment is an important issue, there are other priorities, some of them more pressing, more immediate and more rewarding. In addition, some economic and social hardships of the transition period have pushed environmental problems down the agenda. People were happy that they had been relieved of Communist oppression and, especially, of the economic hardships associated with the old regime. They are eager to adopt the lifestyles of their more affluent neighbours from the Western world. At the same time, the level of environmental education is at a much lower level than in Western Europe or Japan or the USA. This is more than understandable taking into account the general deterioration and underdevelopment of the society prior to 1989. It is imaginable that economic progress will be much more rapid than the enhancement of environmental consciousness and, therefore, it is unfortunately almost sure that this will be accompanied by environmentally unsound consumption habits with inevitable adverse effects.

However, these more than understandable difficulties may have slowed the process of environmental improvement but by no means has it been halted. The basic premises of steady progress are still present. First, an elected government with wide public support, new responsibilities of local governments and communities, the activities of the green NGOs, free access to information and other attributes of an open, democratic society are the most basic prerequisites to substantial environmental improvement. Second, all the relevant elements of the institutional infrastructure of an efficient system of environmental protection are in place. Third, the economic changes have environmentally beneficial consequences. The decrease of production of heavy industry has resulted in an overall reduction of harmful emissions into water and air, and there are other similar effects. Fourth, newly defined property rights have brought responsibility to owners who are made liable for environmental duties. Fifth, the pressure of the new strict air pollution and other laws, together with rather massive investments from public budgets, has already resulted in many finished abatement facilities.

The backbone of the system of environmental protection is the Ministry of Environment of the Czech Republic. This body was established immediately after the November 1989 events and started its operation on 1 January 1990. This was the first ever such body in Czechoslovakia. Fortunately, practically all the responsibilities for environmental protection were given to this Ministry of the Czech Republic which was not much affected by the division of the Czechoslovak Federative Republic in 1992. It could be said that there is continuing development of the activity of the Ministry through the transition period.

In its present form, the whole system of environmental protection is as follows:

- Ministry of Environment in Prague
- Nine territorial offices
- Czech Environmental Inspectorate in Prague
- Regional Inspectorates
- State Environmental Fund
- Czech Environmental Institute (research, information)
- Czech Environmental Agency (nature protection)
- Institutes for nature protection, hydrometeorology, geology, water.

Immediately after its establishment, the Ministry of Environment began an effort to articulate the environmental policy of the new government. During a very short period of time, it produced its 'Rainbow Program'. This was subject to public discussion and subsequently approved by the Government of the Czech Republic and published in early 1991. The Rainbow Program stated that the basic task for the next few years was the accomplishment of the basic economic and political reform of Czech society. This involved a profound conversion of all areas of society in an attempt to dispose of the malignant legacy of more than 40 years of totalitarianism. This conversion, however, did not only have economic and social dimensions but also a significant ecological dimension as well. This implied that from the very beginning, the environmental recovery should be closely connected with the deep restructuring of the whole society.

The success and rate of progress of economic reform will have a decisive influence on the environment. A market economy should enable fast economic growth. There is concern, however, whether this economic development will be sustainable. We understand this term to mean the development of the economy and civilization in a manner that will cover human needs without endangering the quality of the next generation's life and without reducing nature's riches. The objective of economic development is to bring prosperity to all. However, it must be so aimed as not to endanger human health and well-being. It must also not waste natural resources but must save them for the coming generations, and it must maintain the diversity of natural life, which is valuable not only for humans but for life itself.

Modern society has determined that the free application of market forces is the basis for a healthy economy, but these forces must be effectively defined with the aid of laws. The laws must determine the rules and limits of the market system so as not to jeopardize one of the basic human rights – the right to a healthy environment.

The Rainbow Program cautioned that market forces tend to underestimate the real value of natural resources including clean water and air. The market itself often does not recognize this value, which is derived from the uniqueness and irreplaceability of these natural resources. The market economy is also not able accurately to take into consideration long periods of time such as 30 or more years. And, finally, the market system is not capable of attributing adequate value to the objects of nature that seem not to be utilized by anybody; that do not service anybody at the moment and 'only' represent nature's riches.

The establishment of limits and rules governing the impact of market forces reflects the will of the citizens. Every individual person, community and town, as well as the constituents of executive power, i.e. the state administration, all have an irreplaceable responsibility to protect the environment. A particularly important part is played by legislative bodies such as Parliament. In the future, this system will also be supplemented with courts of law.

A primary task is the forming of a new lifestyle. This lifestyle must be based on new ethics that emphasize the qualitative side of human life above the quantitative accommodation of material needs. The right of every person to a healthy environment must be emphasized as well as the obligation of every person to protect the environment. The right of natural organisms to an undisturbed existence must also be emphasized. This approach should also become a basic part of the entire educational process not only for the younger generation but for the whole society. A precondition for public ecological awareness is, of course, free access to all environmental information.

Hazardous factors of a chemical, biological, physical and social nature damage the environment and thus jeopardize human health. These factors have a synergistic effect and are limited to individual components of the environment only in exceptional circumstances. Despite this, the Rainbow Program considered it necessary to improve the condition of the individual components because in this way a more rapid progress could be achieved from the beginning. The following basic objectives were determined:

1. Within two years (by 1993), stop the hitherto unfavourable development of environmental pollution.
2. In the protection of the atmosphere:
   (a) maintain a downward trend in solid emissions;
   (b) stabilize NO_2 emissions at the 1988 level;
   (c) reverse the ascendant development of NO_x emissions, particularly in mobile sources.
3. In the protection of water, achieve a permanent decline from 1990 onwards of the main source of water pollution by reducing pollution from point sources.
4. In the utilization and neutralization of waste:
   (a) by 1992 regulate waste management with laws and through the executive state administration;
   (b) by 1995 create the basic technological conditions for neutralization of hazardous waste and achieve a general reduction of its production.
5. In soil protection:
   (a) in 1991 amend the methodology for anti-erosion protection, and from 1995 onward reduce the extent of soil affected by erosion;
   (b) in 1991 amend the regulations on the use of biocides and other target-aimed substances, define toxic substances that penetrate the food cycle, and by 1993 eliminate their use;
   (c) from 1991 onward eliminate large-area interference with agricultural soil that has a negative impact on the soil and landscape.
6. In forest protection:
   (a) during 1991 propose legal, economic and organizational measures that will minimize the negative effect of outer and inner factors on forest areas.
7. In nature and landscape conservation:
   (a) during 1991 establish a legal base for ensuring territorial systems of ecological stability for the protection of all the components of nature and natural communities and to ensure optimum territorial development under market economy conditions.

Within the indicated time period, these objectives were mostly fulfilled together with detailed ones specified in the document.

Between 1990 and 1995, 14 new Acts of law, numerous amendments and dozens of other legal provisions were adopted and enacted to establish a system of normative, economic, institutional and informational instruments to protect the environment in the Czech Republic.

Act No. 17/1992 Sb, on environment protection, provides the basic notions and principles related to the protection of the environment. The Act is a framework law that is formulated in general terms.

Obligations relating to environmental impact assessment are provided in the Act of the Czech National Council No. 244/1992 Sb. The law provides obligations and procedures for buildings and structures but provisions relating to the assessment of products, concepts and transboundary impacts are not clearly defined; provisions relating to the binding nature of the law and certain elements of procedure are not provided.

Act of the Czech National Council No. 282/1992 Sb, regarding the responsibilities of the Czech Environmental Inspection and its jurisdiction in the protection of forest areas, to a large degree unifies the control activities in the sphere of the environment. Act of the Czech National Council No. 388/1991 Sb, regarding the State Environment Fund, unifies state funding of environmental protection. Both laws provide an adequate legal framework for those activities and responsibilities associated with the operation of these respective institutions.

The protection of air is provided in Act No. 309/1991 Sb, regarding the protection of the air from pollutants (amended by Act No. 211/1994 Sb), in the follow-up Act No. 389/1991 Sb, regarding obligations for state administration of air protection and provisions for air pollution charges (amended by Act No. 212/1994 Sb) and in other legal provisions. The laws provide legal provisions to stimulate the industrial sector to take remedial measures and undertake investments to make environmental improvements (these provisions have resulted in investments in the order of dozens of billions of Czech crowns). It is also anticipated that the compliance schedules provided in these laws will result in fundamental improvements in air quality by requiring stringent reductions in emissions of harmful substances. Normative instruments (emission and immission limits) have been adequately established in these laws. However, the provisions concerning the air pollution charges only fulfil their fiscal function to a limited extent and as currently structured are inadequate in providing sufficient incentives to industry to reduce emissions and/or introduce pollution control technologies. Inadequacies in the structure of air pollution charges include low charge rates, difficulties in adjusting these rates to account for higher costs and limited possibilities for suspending the pollution charges. The 1998 deadline for compliance with the prescribed emission limits is projected to be met by approximately 75% of polluters.

The protection of the Earth's ozone layer is governed by the recent Act No. 86/1995 Sb. This Act fully complies with international commitments for ozone protection which the Czech Republic has endorsed or will be endorsing within a prescribed timetable.

The protection of water and the provisions related to water management are provided in Act No. 138/1973 Sb and follow-up provisions. The legal norms provided in this Act essentially fulfil its mandate for water protection and management as is evidenced by systematic improvements in water quality. However, this Act does not adequately define existing property rights and ownership issues with regard to waters and their environment. The provisions regarding the use of economic instruments (charges for wastewater discharges) fulfil their fiscal and incentive roles to a limited extent. The new water law is being prepared.

Soil protection is governed by Act No. 334/1992 Sb, regarding the protection of the agricultural soil fund. This law concerns soil as a means of production but does not provide measures to address the environmental issues related to soil protection. In addition, in some instances, this Act has hindered construction in urban areas.

Nature protection is governed by Act No. 114/1992 Sb, on nature and landscape protection and the follow-up provisions. Excluding some minor points, this Act does not apply economic instruments to encourage protection and preservation of nature. The creation of the Czech Republic's national parks is provided by government decrees. Accepted international commitments (e.g. CITES) are not endorsed under this legislation but the corresponding laws are under preparation.

The present legal norms regarding waste management proceed from Act No. 238/1991 Sb, on waste management and the follow-up provisions. At this time, these laws are insufficient and do not address the current situation nor do they comply with the relevant standards issued by the EU and OECD. In addition, these laws do not provide adequate incentives to encourage waste minimization (especially in the area of hazardous waste). Existing economic instruments, such as charges for waste disposal, often encourage undesirable types of waste disposal (unsorted land filling) and do not stimulate waste producers to collect, salvage, sort and recycle the waste. Many waste producers are unable to make the necessary investments in environmental technologies and their inability to make these investments is further hampered by their financial obligation to pay waste disposal charges. In addition, legislation regarding the transboundary movement of secondary raw materials is excessively stringent and incompatible with European standards.

The existing legislation regarding the management of chemical substances and preparations is provided by Act No. 20/1966 Sb, on public health and follow-up legislation. This Act of law inadequately defines properties of chemicals and chemical preparations with regard to environmental protection and the appropriate procedures for handling chemical substances which are hazardous to human health and the environment. Current legislation in this area does not provide necessary protection measures concerning the transboundary movement of chemical substances.

The current system of environmental legislation in the Czech Republic covers the most important areas of environmental protection more or less effectively. There are deficiencies in providing uniform terminology and there are no provisions to address packaging requirements, the regulation of genetically modified organisms or accident prevention and response. In addition, the provisions addressing the remediation of old environmentally damaged sites are inadequate.

However, an expedient and economically effective instrument has been the negotiation of voluntary agreements between the state administration and pollution producers. Voluntary agreements have proven to be advantageous in cases where there are a limited number of producers of a specific type of pollution and/or in those cases where the individual conditions provide mutual economic and environmental benefits for the State and the polluters. In 1995, the Ministry of the Environment of the Czech Republic concluded such an agreement with the Association of Manufacturers of Washing Powders for the gradual reduction of environmentally harmful substances in their products.

The rules for management of all kinds of forests are provided by Act No. 289/1995 Sb. Here, the rights and obligations of all forest owners are specified. The State owns about 50% of forests, the rest being in private hands. The role of the State in forest management is still crucial. The rights of owners are limited.

The Government did not consider it necessary to work out a new version of a programmatic policy document until August 1995 when the State Environmental Policy was approved by the Cabinet of Ministers. This material is a relatively brief text that gives a brief analysis of the state and development of the environment in the Czech Republic after 1989. It spells out fundamentals, principles and priorities of environmental policy. The fundamental postulate which guides the formulation of environmental policy in the Czech Republic is the responsibility of the present generation to preserve and transmit fundamental life values to future generations (clean air and water, productive land, healthy foods, a safe climate).

State Environmental Policy of the Czech Republic, therefore, stresses the rational and efficient use of resources. Emphasis is placed on recycling, limiting pollution (to a level which does not produce irreversible damage to human health and/or nature), respecting the importance of biological diversity and seeking economically favourable ways of meeting man's basic needs without jeopardizing environmental systems.

The State Environmental Policy complies with internationally accepted principles for environmental protection and as such was guided by the Czech Republic's efforts to join the OECD and the European Union (Czech Republic joined OECD in Dec. 1995). Priority measures will continue to be developed to resolve particularly difficult local and regional problems and to meet commitments associated with international agreements and conventions to which the Czech Republic has acceded or intends to accede.

The Czech Republic's environmental policy is conceived as a dynamic approach which will facilitate identifying ecologically, economically, socially and politically optimal policies as opposed to establishing an inflexible system which could hamper economic development and lead to state control.

The goals of environmental policy and the instruments for their attainment must be formulated to maximize the potential for creating an optimal system. This process proceeds from finding a socially acceptable level of environmental and health risks. Such an approach must recognize that striving for a 'zero risk' level (the absolute and complete elimination of pollution and harmful factors) is not always necessary to attain environmental protection goals and is cost-prohibitive. The system of policy instruments which is selected needs to be environmentally and economically acceptable and integrated with the social, political, regional and international aspects. Therefore, the goals of environmental policy and the instruments for their attainment cannot be assessed separately and must be developed so as to integrate with other goals of the State in related sectors. The measurement of the importance of these individual goals is the result of public choice.

The environmental policy of the Czech Republic proceeds from the conviction that certain activities relating to the protection of the environment which fall outside of the State's irreplaceable responsibility to provide and protect non-market environmental goods may be more efficiently implemented by the private sector in conjunction with the respective legal and institutional framework of the State.

The irreplaceable responsibility of the State includes those activities related to environmental protection which cannot be provided by other entities. These areas mainly concern the protection of those components of the environment which cannot be privately owned, or whose use may not be delimited (the air, the climate, important watercourses, ecosystems), as well as the development of a legal framework to protect the environment and guarantee the fulfilment of international commitments.

It is therefore desirable, whenever possible, to transfer the costs associated with environmental improvements from the public to the private sector. This process will necessitate that financial flows from the State are transferred to financial flows between private persons, municipalities and localities.

The State Environmental Policy of the Czech Republic is guided by standard principles which are applied in the development of normative and economic instruments. The critical load principle, the concept of technical availability and the precautionary principle are applied in the development of normative instruments (limits, standards, regulations, bans, deadlines).

The economic instruments of environmental policy (charges, tax measures, subsidies) are derived from the principle of individual responsibility for the costs of environmental damage. The principle is based on the assumption that firms and individuals should be explicitly faced with the costs of environmental damage resulting from their activities. In the *short-term context*, the basic criteria for determining priorities are the reduction of health hazards and the protection of those components of the environment which may transmit pollution to the other components. The order of priorities established for 1995–98 is as follows:

1. Improving air quality through the reduction of harmful emissions.
2. Improving water quality by limiting pollution discharges.
3. Reducing the production of wastes (namely, hazardous waste).
4. Eliminating the impacts of harmful physical and chemical factors. The main objective is to reduce the percentage of the population which is exposed to increased risk factors such as noise, radon, toxic chemicals and preparations.
5. Remedying previous environmental damage which poses acute hazards for human health and the environment.

In the *medium-term context* (1999–2005), projections indicate that a significant portion of the above-mentioned priority problems will be, at least partially, resolved and, therefore, resolution of the following problems will increase in importance:

6. Creating land use provisions which will safeguard the efficient protection of the individual components of the environment and fulfil international commitments through regional planning.
7. Increasing the water retention capacity of land by improving the revitalizing measures.
8. Continuing the reconstruction of forest growth in areas damaged by air pollution.
9. Continuing the reclamation of areas devastated by mining activities.

The *long-term* or, more precisely, perpetual priority areas of the State's Environmental Policy are:

10. Climate protection by reducing greenhouse gas emissions by changing the structure of energy sources, reducing energy demand, promoting energy savings and taking measures to increase capacity of sinks.
11. Protection of the Earth's ozone layer by continuing the gradual phasing out of ozone-depleting substances.
12. Protection of biological diversity primarily by minimizing harmful impacts, revitalizing biotopes, and protecting and reintroducing endangered indigenous species.

The State Environmental Policy concludes that the implementation of the approved policies will create the necessary conditions to ensure that by the year 2005, the quality of the environment in the Czech Republic will be comparable to levels achieved in the early 1990s by Western countries, and that favourable conditions will have been created for the systematic improvement of the state of the environment.

Similarly, as the Rainbow Program several years ago, the new State Environmental Policy is rather modest. It is quite realistic to expect that its objectives will be fulfilled and that the state of the Czech environment will be improved markedly.

The generally effective work of the Ministry of Environment and Government as well as the positive attitude of the Parliament is based on the continued public support. Even if the initial environmental euphoria of the early phases of the transition period is long gone the interest of Czech citizens in the environment is solidly present, e.g. according to several recent public opinion polls, people regard the global environmental problems as the most important ones, more important than a global war, new illnesses or North–South gap. Environmental pollution is believed to be the main cause of people's ill-health.

There are several dozens of environmental non-governmental organizations, practically all of them were created after November 1989. Some of them are focused primarily on nature conservation such as the Czech Union of Nature Conservation. The Union is the largest of green NGOs of the Czech Republic having about 10 000 members. It has its local organizations in most of the Czech districts and has created dozens of environmental education centres. Many organizations are focused on children and youth, some of them are not specialized in environment or in nature protection but have strong interests in this direction like the organization Junák – the Union of Czech boy and girl scouts.

Many NGOs are specialized in different fields of environment and are trying to lobby the Parliament regarding new environmental and other laws. Among them, the following have prominent positions:

- Duha – Friends of the Earth
- Children of the Earth
- Brontosaurus
- Tereza
- Koniklec.

There also exists an umbrella organization called Zelený kruh (Green Circle). In the Czech Republic, several international green organizations operate, e.g. Greenpeace.

Recently, the partnership among environmentalists and the private sector has been recognized as an essential prerequisite for effective environmental protection. Pioneering this partnership is the Czech Environmental Management Centre which also acts as a secretariat to the Czech Business Council for Sustainable Development.

Also important are some of the expert non-profit organizations specializing in various aspects of environment-related projects. Highly successful are, among others, SEVEn, the Centre for Energy Efficiency, and the Institute for Environmental Policy.

It is generally accepted that environmental education has a crucial role for any lasting success of environmental protection. Many institutions at different levels are active in this direction, starting with the Government, which issued the special degree fostering environmental education in 1992, the Ministry of Education and numerous environmental NGOs. There are many grants supporting environmental education, given by various donors including international and intergovernmental organizations like the European Union.

REFERENCES

Ministry of Environment of the Czech Republic, Czechoslovak Academy of Sciences (1990) *Environment of the Czech Republic*. Academia, Prague. English translation. Ekocentrum Brno, 1991, 315pp.

Ministry of Environment of the Czech Republic (1991) *Environmental Recovery Programme for the Czech Republic*. Academia, Prague, 93pp.

Ministry of Environment of the Czech Republic (1993, 1994) *Environmental Year-Books for 1991, 1992, 1993, 1994*. Czech Ecological Institute, Prague.

Ministry of Environment of the Czech Republic (1994) *First National Communication for the Framework Convention on Climate Change*. Prague, 41pp.

Ministry of Environment of the Czech Republic (1995) *State Environmental Policy*. Prague, 15pp.

Ministry of Environment of the Czech Republic (1996) *Report on the State of the Environment of the Czech Republic in 1994*. Prague, 93pp.

Moldan, B. (1995) *Environmental Cost Internalization in the Commodity Sector. A Case Study for the Czech Republic* (mimeograph). UNCTAD, Geneva, 41pp.

CHAPTER 4

Hungary

Zsuzsa Lehoczki and Zsuzsanna Balogh

Despite the situation we inherited, a historical opportunity has opened for environmental protection. Several environmental initiatives and groups were among the forces that gave decisive push to the process of democratic change.

(*Hungary's National Renewal Program* (1990–1992) Budapest, 1990, p. 81.)

Bride: a woman who has a bright future behind her.

(Ambrose Bierce: *Devil's Dictionary*
(as quoted in Gerald Durrell: *Marrying off Mother*).)

4.1 INTRODUCTION

The first government of the democratic Hungarian Republic placed high priority on re-establishing environmental policy. Enthusiastic promises and commitments in the Government's programme suggested a bright future for environmental protection. This was supported by expectations about political, social and economic changes. Environmentalists had high hopes since environmental groups were active in preparing the stage for the peaceful revolution of 1989.

Environmental problems, particularly the ones associated with the Bös-Nagymaros Barrage System Construction Project, were issues around which opposition to the old socialist system was organized. They became good focal points for different forces and groups to systematically attack the centralized political system. Protests and marches against the Bös-Nagymaros Dam construction calling for saving environmental values were instrumental in unifying political opposition. Moreover, discontinued construction of the Dam was one of the first clear signals that the ruling 'old' Socialist Party was willing to go along with requested economic and political reforms. Therefore, environmentalists had optimistic expectations and they envisioned ambitious plans for environmental protection and nature conservation.

The Environmental Challenge for Central European Economies in Transition.
Edited by J. Klarer and B. Moldan. © 1997 John Wiley & Sons Ltd.

High hopes certainly have been less than fully realized in the longer than expected years of transition. When the multi-party system was established, many leaders of the environmental groups became leading political figures in different parties and their concerns were 'broadened' beyond environmental issues. In fact, the need to address environmental problems started to slide down the political agenda.

The initial 'political credit' was barely translated into professional strengthening of the environmental administration. In 1990, redefinition of the organizational set-up, tasks and responsibilities for environmental administration were far too dominated by the issues of separating the environmental protection administration from water management. Ambitions for constraining the powerful 'water lobby' often overshadowed requirements for building an efficient network of organizations for environmental protection administration.

Thus, we could observe the process of the initial optimism and political inertia fading and giving way to the day-to-day drudgery of environmental policy-making. In 1990, during the discussion of the National Renewal Programme, there was enthusiastic support for the fairly vague environmental section of the programme. The same Parliament, however, never discussed a new environmental protection Act. In the first phase of privatization, environmental concerns and liability questions were ignored. These are just two important examples out of several others showing that real and strong constraints for environmental policy quickly emerged in the period of transition.

The aim of this chapter is to provide a description and assessment of the changes in environmental protection. We will provide an account of underlying trends and causes and indicate how the high hopes for environmental protection have been hit hard by economic and social constraints. Based on available economic and environmental data and personal experiences, we will attempt to identify key factors and issues which will shape the state of the environment in Hungary in the future.

4.2 PRESENT TRENDS IN THE STATE OF THE ENVIRONMENT

A five-year period is not long enough to trace long-term trends about environmental impacts of the emerging market economy and democratic society with a high level of certainty. The slowly evolving process of economic restructuring makes it particularly difficult to identify long-lasting impacts, but some tendencies can be traced and preliminary assessment can be made on the environmental consequences of the transition. Selected economic indicators for the period 1990–94 are presented in Table 4.1.

The overall tendency is a clear decline in emission, in the volume of water effluent, and agricultural non-point source pollution. Correspondingly, there is general improvement in air and surface water quality, but both local and temporary pollution concentrations still occur. Trends in soil quality indicate

Table 4.1   Selected economic indicators of Hungary, 1990–94

	1990	1991	1992	1993	1994
GDP growth (%)	−3.5	−11.9	−3.0	−0.8	2.0
Private consumption growth (%)	−3.6	−5.8	−0.5	1.3	1.3
Investment growth (%)	−8.3	−0.6	−2.7	1.7	11.5
Wage index (1990=100)	100.0	96.0	97.5	96.5	99.0
Inflation (CPI, annual average)	28.9	35.0	23.0	22.5	18.8
As percentage of GDP:					
Current account balance (%)	1.0	1.2	1.0	−9.0	−9.5
General government balance (%)	0.5	−2.1	−5.4	−6.6	−6.4
Net savings of households (%)	5.3	9.4	7.9	3.0	4.8

Source: World Bank (1995).

a more complicated picture since accumulated past pollution causes a bigger problem there.

The important question with respect to the rather positive picture is to what extent is this simply due to severe economic recession, particularly to the sharp decline in industrial and agricultural output. If economic recession is the major factor, then environmental improvement is only a temporary relief. Longer lasting impacts can be expected from structural changes in the economy driven by price mechanisms that should result in more efficient resource uses. Unfortunately, available data do not allow separation of the two possible forces: recession and structural changes. First, the environmental database is rather disorganized and unreliable. Second, it is rather difficult to link the environmental database to economic databases.

Air pollution

There is little comprehensive analysis available on the changes and tendencies in ambient air quality in the period of transition. Average concentrations of most pollutants are below standards in the bigger settlements. The high variation in the concentration values indicates some temporary 'hot spots' in some cities.

There are clear indications, however, that air quality in some of the traditional 'hot spots' has substantially improved. These are the old socialist industrial centres that are most affected by restructuring and industrial recession. There is no consensus, however, whether air quality would disappear quickly with increased industrial production or be maintained through successful and efficient industrial restructuring.

One tendency, on the other hand, is very clear – the increase in urban traffic is related to air pollution. Large increases in the private vehicle fleet, particularly the large import of old Western cars, accompanied by declining

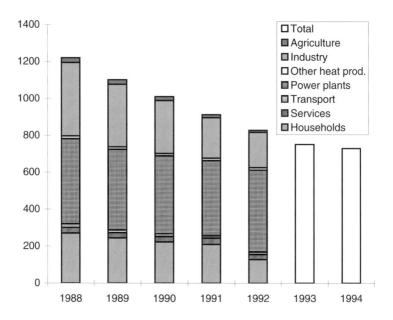

Figure 4.1 SO_2 emissions by sectors in Hungary (kt a^{-1}), 1988–94. *Source*: Ministry of Environment (1994). Estimates for 1993 and 1994: Tajthy (1995)

public transport services is responsible for air pollution along busy roads and streets in urban areas, particularly in Budapest.

Emissions of the major air pollutants have steadily declined since the mid-1980s. Figures 4.1–4.3 and Tables 4.2–4.3 show emissions of five pollutants: SO_2, NO_x, particulates, ozone depleting substances (ODS) and volatile organic compounds (VOC). The pace of decline varies across these pollutants. SO_2 and NO_x emissions decline fairly steadily resulting in an overall decrease of 25% between 1988 and 1992. Particulate emission has a sharper decline between 1988 and 1989 due to an investment programme, but then it follows a slower but still steadily declining pattern as well. Falling emission, in the case of these three pollutants, is mostly due to declining industrial output.

VOC and ODS emission decline is large. It is about 34% between 1988 and 1992 for VOC, and 50–80% for different categories of ODS. This sharp VOC and ODS emission reduction can be attributed to environmental policy measures. It is, in substantial part, due to technology changes that have been introduced as responses to environmental regulation on reducing or phasing out the use of products responsible for VOC or ODS emissions.

The power sector has the largest and continuously growing share in SO_2 release. The production level of the sector is slightly increasing since reduced electricity demand has reduced export rather than affecting domestic production. That stabilizes SO_2 emission from the power sector. Meantime, industrial

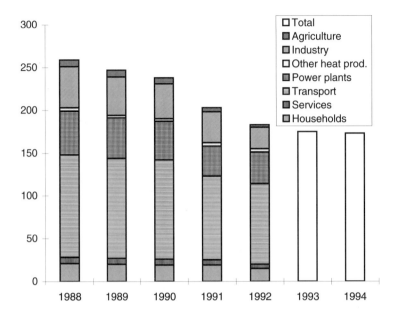

Figure 4.2   NO$_x$ emissions by sector in Hungary (kt a^{-1}), 1988–94. *Source*: Ministry of Environment (1994). Estimates for 1993 and 1994: Tajthy (1995)

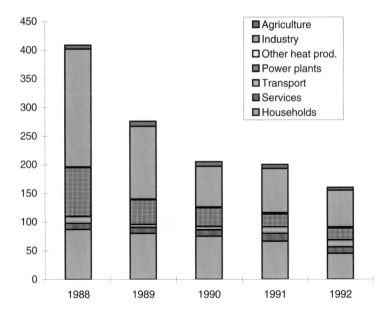

Figure 4.3   Particulate emissions by sectors in Hungary (kt a^{-1}), 1988–92. *Source*: Ministry of Environment (1994)

Table 4.2   ODS emissions in Hungary, 1990–92 (t a^{-1})

	1990	1991	1992
I. category	4410	2300	1696
II. category	734	250	141
III. category		1100	647
out of which HCFC		na	319

Source: Ministry of Environment (1994).

Table 4.3   VOC emissions in Hungary, 1989–92 (kt a^{-1})

	1989	1991	1992
Energy sector	1	1	1
Transport	91	73	68
Oil production and refineries	35	25	24
Solvents and paints	79	45	43
TOTAL	205	144	136

Source: Ministry of Environment (1994).

emission is declining due to output decline. Household emissions decreased by 50% mostly due to an extended natural gas network development programme. The length of the gas pipe network doubled between 1990 and 1994, from 22 559 km to 45 325 km, and the quantity of piped gas sold to households increased from 1862 million m^3 to 2857 million m^3, during the same period (KSH, 1995). Further decline may be hindered by the substantial increase in the natural gas price that may discourage conversions to natural gas. The price of natural gas for households basically was kept at the same level between 1992 and 1994. Meantime, the price of other energy sources for residential use increased, thus creating a relative price advantage for natural gas. At the end of 1994, the Government adapted the proposal for an energy price increase including a 53% increase of end use gas prices for households by 1 January 1995 (IEA, 1995).

The NO$_x$ emission decline is the result of reduced emission from the four sectors that are major sources. We must note, however, that the decline in NO$_x$ is rather uneven. The moderate decrease in 1988 and 1989 was followed by sharp decline in 1990, most likely due to the first drastic increase in gasoline prices. The annual average price of gasoline was about 26 HUF l^{-1} (US$ 0.52) in 1988 and 1989 (HUF = Hungarian forints). It increased to 45.1 HUF l^{-1} (US$ 0.71) in 1990 and 57.4 HUF l^{-1} (US$ 0.77) in 1991 (Papp, 1995). In 1991 there again was only moderate decline. The modest decline can be explained as reduced economic activity of the transport sector which caused reductions in NO$_x$ emissions. However, between 1992 and 1994 the transport sector showed no further decline (KSH, 1995), and air pollution from the sector has been stabilized as well. Transport related emission has decreased at a somewhat slower pace than emissions from other sectors. Therefore, its share has grown.

We can trace similar tendencies for VOC emissions since the largest contributor is the transport sector. The other big pollution source, the solvent and paint industry, reduced its pollution through developing alternative products. In November 1991, the Government of Hungary signed and ratified an international protocol in the frame of the Convention on Long-Range Transboundary Air Pollution that requires maintenance of VOC emissions at their 1988 level by 1999. Due to the longer than expected recession, this target is not likely to be difficult to achieve. In fact, achieving more ambitious targets can be feasible, as well as some control measures in the transport sector.

Emission decline was the most substantial for ODS. Since Hungary has signed the Montreal Protocol, a new regulation was passed to fulfil the international obligation. The regulation contains a phasing-out schedule and stipulates fines as an enforcement tool. Another step in enforcement is the newly introduced product charge on refrigerators and refrigerants. The product charge raises funds for supporting the phase-out through subsidies.

Trends in energy use

Wasteful use of resources was one important feature of the centrally planned economies. This was often characterized by the high value for energy use per unit of GDP. There were clear expectations that establishing market prices for the resources, such as energy, would result in improved efficiency.

The total energy use fluctuated in the last few years as shown in Tables 4.4 and 4.5. This is mostly due to different time patterns of the 'restructuring process' in industrial, transport and household consumption. In industry and transport, the total energy use sharply decreased in 1991 due to a decline in activity and the shutting down of plants. After 1991, the industrial decline continued while transport sector performance stagnated. Total energy use in these sectors decreased slightly until 1993. From 1990 to 1991, household energy consumption increased in spite of increasing prices. This is possibly due to the increased availability of household electric equipment and a temporary demand for energy. In the short run, there are few possibilities to respond to price increase with improved household energy savings and efficiency improvements.

The ratio of coal in the structure of primary energy sources is reduced and that of natural gas is increasing. Even though domestic coal mining is subsidized through a special arrangement between coal mines and power plant, orchestrated by the Government, the unions agreed to a gradual closing schedule for most of the mines by 1997 (IEA, 1995, pp. 114–116). This will end the practice of subsidizing coal use and its adverse environmental impact.

Table 4.4   Energy production, 1990–94 (PJ a^{-1})

	1990	1992	1993	1994
Coal	188	152	139	136
Oil	78	75	70	74
Diesel oil	16	16	16	16
Natural gas	160	155	169	163
PB gas from mining	10	10	10	10
Hydropower generation	2	2	2	2
Nuclear power generat.	137	140	136	136
Fire wood	12	13	13	13
Others			8	7
TOTAL	603	560	562	557

Source: IEA (1995).

Table 4.5   Energy consumption in Hungary, 19990–94 (PJ a^{-1})

	1990	1992	1993	1994
Coal	188	203	183	178
Oil	78	305	305	319
Diesel oil	16	58	86	74
Natural gas	160	327	353	368
PB gas from mining	10	10	10	10
Hydropower generation	2	36	27	27
Nuclear power generat.	137	140	136	136
Fire wood	12	13	13	13
Others	0	0	8	7
Import–Export	654	−51	−64	−50
TOTAL	1257	1041	1056	1082

Source: IEA (1995).

The key issues in the energy sector development are as follows:

–   Prediction of future demand.
–   Assessment of the need for a new power plant: nuclear or fossil fuel based.
–   Determination of how energy (gasoline, electricity) should be priced.

The decision on the need for a new base power plant has been postponed until after the year 2000. Delaying the decision is due partly to the large variations in projected demand, particularly in the industrial sector, and partly to the uncertainties associated with privatization in the energy sector.

The large investment cost of a new power plant has created some interests in demand side management and energy efficiency improvements. The crucial question for environmental protection, however, is the still unknown impact of industrial restructuring on the energy intensity of the economy.

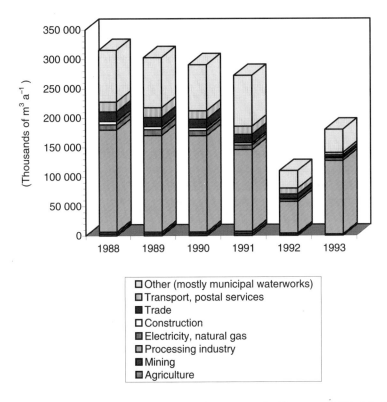

Figure 4.4   Wastewater discharges into surface waters in Hungary, 1988–93. *Source*:
KSH (1995)

Water pollution

Water quality protection has improved because of economic recession as well,
but comprehensive, up-to-date assessment is lacking. Effluent discharge to
surface water has decreased continuously since 1988 as shown by Figure 4.4.
Unfortunately, precise analysis of the trend is not possible because the
statistical classification system was changed in 1992 and wastewater reporting
was not done properly. Even though the 1993 data seem to approach the
'normal' trend, we still must calculate with some uncertainties in our database.
    The number of wastewater discharge sources is volatile even beyond the
statistical classification problem. The number of industrial sources seems to be
stabilized thereby indicating an extended number of bankruptcy procedures
and/or increasing number of small, non-reporting enterprises. Agricultural
sources have a sharply declining share in effluents due to severe organizational
problems and low production levels in the sector. The main question is, just as
in many other cases, whether some increase in the sector's output would cause

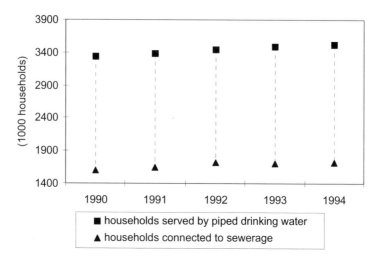

Figure 4.5   Number of households connected to public water utility in Hungary, 1900–94. *Source*: KSH (1995)

disproportionately higher pollution due to lack of resources for better production methods and higher quality input resources.

The municipal sector is particularly difficult to describe due to constant reorganization of water works and wastewater treatment facilities. Establishing full cost pricing and privatization processes has created a long-lasting transition for the sector. Household wastewater discharges continue to be a problem for both surface and underground water pollution. About 90% of the households are connected to a piped drinking water supply, but less than half of those households are connected to a sewage system as shown in Figure 4.5.

Even though water consumption has decreased since water prices increased (see Table 4.6), unconnected households are potential non-point-source polluters. There are standards developed for septic tanks, and there is a subsidy programme for municipalities for providing collection and treatment of effluents from septic tanks. The regulation, however, is weakly enforced and companies rarely apply for the subsidy because it is usually more profitable to provide the service as part of the 'grey economy'. The municipal sector often pollutes surface water. Only 50% of the sewage is treated, and only 33% of the treated wastewater receives biological treatment (see Table 4.7). Of the untreated wastewater from the sewage system, 63% originates from the capital and the five biggest cities of the country.

Waste management

Data concerning municipal solid waste are rather unreliable. Collected and transported household waste has declined since 1988 with a particularly sharp

Table 4.6   Quantity of water sold in Hungary (million m³)

	1980	1990	1992	1993
Drinking water	793	889	841	809
Water for industrial use	64	66	63	60
Water for agricultural use	530	819	543	514
TOTAL	1387	1774	1447	1384

Source: KSH (1995).

Table 4.7   Wastewater treatment in Hungary (million m³)

	1980	1990	1991	1992	1993
Waste water requiring treatment	951	970	967	850	802
properly treated	143	206	272	231	167
partly treated	626	666	595	540	548
without treatment	183	98	100	79	87
Used water not requiring treatment	1259	3889	5227	2082	1963

Source: KSH (1995).

decrease in 1990 (see Table 4.8). There is little information available to explain the change but possible causes range from the following changes:

– Households have become poorer and produce less waste.
– Due to organizational and financial uncertainties there is less waste collection and transport offered to households.
– The rising prices of services reduce household willingness to subscribe.

The number of disposal facilities has been increasing since 1990. It was 333 in 1993. Municipal solid waste management has not reached high quality. More than half of the disposal sites do not comply with environmental regulation. According to experts' estimates about 70% of the municipal solid waste is disposed of, about 8% is incinerated and only about 2% is recycled. There is no information on the rest (about 20%). It most likely ends up as illegal littering.

Municipal solid waste management was among the many services the Socialist State promised to provide free of charge for citizens. Now, in the midst of starting to price so many public services, it is politically rather difficult to charge proper fees for waste disposal and management. Cost recovery and financing questions, which are subject to a lot of debate even in the market economies, add to the problems of transition. Policies to encourage recycling and proper 'transition' financing measures are critical for improving solid waste management.

Table 4.8 Municipal solid waste in Hungary, selected indicators

	1980	1985	1989	1990	1991	1992	1993	1994
Collected and transported municipal solid waste (1000 m³)	9 846	13 791	18 283	16 685	16 092	16 790	16 403	17 153
Treated solid waste (1000 m³)	–	–	–	16 372	15 890	16 780	16 394	–
Number of disposal sites	–	–	–	299	294	315	333	414
Ratio of dwellings included in the waste collection system (%)	–	–	–	65	–	66	67	70

Source: KSH (1995); authors' enquiry at Ministry of Environment and Regional Policy.

Table 4.9   The volume of hazardous waste in Hungary
($\times$ 1000 t a^{-1})

	1989	1990	1992
Agriculture	417	297	168
Mining	92	77	146
Processing industry	2007	1951	1435
Power sector	138	124	186
Construction	58	66	23
Trade	73	59	91
Transport and postal services	54	57	42
Others	108	109	152
Not identifiable	32	27	–
TOTAL	2979	2767	2243

Source: Authors' enquiry (Hazardous Waste Data Base of Ministry of Environment and Regional Policy).

Data on hazardous waste generation and disposal are even more scattered and unreliable. There is an official database compiled from self-reports in connection with generating, transporting and handling hazardous waste. Since these self-reports are the basis for assessing hazardous waste fines, they are likely downward-biased with respect to generation and upward-biased with respect to treatment. There are initiatives to improve this database with the help of expert judgements and a material balance approach. Meantime it is necessary to compile an inventory of the illegal hazardous waste dumpsites. These initiatives have not yet yielded results, therefore, we must confine our analysis to the existing database.

The total volume of hazardous waste has decreased by 25% between 1989 and 1992. Detailed data in Table 4.9 show that the lower volume of hazardous waste generation basically reflects a sharp decline in waste generation in the processing industry and agriculture. The share of the processing industry has decreased from 67% in 1989 to 64% in 1992. Declining production in these sectors is the most likely cause for that decrease. The assertion that economic recession is the most important explanation in the declining trend is supported by the observation that where the share of the power sector's output level is slightly increasing, waste generation also is increasing. This raises the power sector's share.

The number of sources generating hazardous waste is fairly stable. In the number of polluting sources, the share of processing industry is only about 27–29%, indicating large volumes of hazardous waste generation at individual sources. This may suggest that new, smaller industrial sources do not report in order to avoid the fine. Meantime, relatively large numbers of agricultural and service sector sources generate small volumes individually. The share of the most harmful waste is fairly low, about 5–6%. The processing industry is responsible for a large part (68%) of waste with an average hazard level.

Table 4.10   Destination of hazardous waste generated in Hungary, 1989–92
($\times$ 1000 t a^{-1})

	1989	1990	1991	1992
Waste generated	2863	2701	2516	2666
Accumulated inside the plants generating the waste	484	678	1108	1138
Temporarily stored inside the plants	95	236	248	169
Neutralized inside the plants	1385	1131	920	1101
Recycled inside the plants	54	46	61	95
Treated inside the plants	1533	1413	1228	1365
Transported out of the plants	1185	1166	1027	1115
Temporarily stored outside of the plants	301	198	138	160
Neutralized outside of the plants	184	148	109	125
Recycled outside of the plants	64	57	19	13
Passed to other 'treatment organizations'	637	763	760	818

Source: Authors' enquiry (Hazardous Waste Data Base of Ministry of Environment and Regional Policy).

Data on hazardous waste treatment do not indicate a favourable picture. The reported waste generation exceeds the reported treatment each year. This is partly the result of an article in the hazardous waste legislation that allows temporary storage for an unlimited time with little specification on the precise characteristics of temporary storage. Since 1992 the accumulated hazardous waste has grown to 1.14 million tonnes.

Half of the treatments take place inside the plants generating the waste and the other half outside (see Table 4.10). The ratio of waste disposed of in temporary storage is about 13% each year. The ratio of passing hazardous waste to other treatment organizations is high, 33%. This most often means starting a route at the end of which is illegal disposal or simply accumulation of the waste. This assertion is supported by the very low volume of waste passed to the 'other' sector which contains professional licensed hazardous waste treatment facilities. Meantime, there are several bankruptcy cases against companies 'specialized' at taking hazardous waste and with the liability to treat it. Usually, the treatment cost of their accumulated waste is several magnitudes higher than the value of their assets, therefore, effective liability for clean-up cannot be assigned to these companies.

Nature conservation

In Hungary about 40 plant species and 53 animal species have become extinct, and about 1130 more species are threatened. There are several protected areas in Hungary. These cover about 5188 km^2, nearly 6% of the country's area. They include all the major habitat types, the largest part of which are forested areas. All 2245 caves are protected. Of these, 97 are intensively protected.

Wildlife is still fairly abundant, but excessive hunting and expansion of agriculture and forestry threatens them.

Three national parks (Aggtelek, Hortobágy, Kiskunság) and two landscape protection areas (Pilis and Lake Fertő) were designated as biosphere reserves in 1979 and 1990. There are also several designated Ramsar sites (Hortobágy, wetlands on the Tisza, Kis-Balaton).

The major task is to protect natural values and resources in the period of massive ownership changes. There has been extensive discussion on how threatening private property is to natural resources such as forests or protected areas. Uncertainties in property right assignments and the transition from state ownership to clearly defined private ownership invites exploitation of those resources. For example, the question was raised as to whether protected areas could be claimed back by their former private owners (those who owned them before nationalization of the land in the 1950s) or whether the property could be used for land compensation schemes. Due to a legislative error, re-privatization of protected areas was allowed, but a few months later this decision was ruled as unconstitutional. By then, however, large areas were privatized. They now must be bought back to state ownership according to the ruling of the Constitutional Court. Scarce budgetary resources make the process rather slow. In the meantime large forestry cuttings and log exports to Western countries are reported.

4.3 INSTITUTIONS AND POLICIES: CHANGES IN THE PUBLIC AND PRIVATE SECTORS

Changes in governmental tasks and responsibilities

In 1990, a new organizational setting was created for environmental administration. Environmental protection was separated from water management. The justification was that it was necessary to draw the line clearly between the utilization of water resources and the environmental supervision of that utilization. In the new administration, environmental protection has been coupled with regional planning while water management has been shifted to the Ministry of Transport, Telecommunication and Water Management. Subsequently, the environmental and water authorities were separated into 12 environmental inspectorates and 12 water authorities.

In the revised governmental structure, the *Minister of Environment and Regional Policy* has the main responsibility for co-ordinating and carrying out environmental protection. (The Minister also has responsibility for regional planning and building regulation.) In the field of environment, the scope of tasks covers:

- Developing an overall environmental strategy
- Analysing and evaluating the state of the environment

- Running and/or co-ordinating the appropriate monitoring system and information network
- Developing the system and methodology for Environmental Impact Assessments
- Encouraging environmentally-friendly practices.

In the scope of air quality protection, the Minister:

- Defines the rules and methods for setting emission standards
- Develops rules and methodology for observing, measuring and analysing transmission of air pollutants
- Defines regulations and requirements for air pollution abatement
- Takes part in setting ambient air quality standards.

In the scope of water quality and quantity protection, the Minister:

- Defines the water qualification system
- Defines the environmental conditions and requirements for interventions into the water resources and enforces them
- Defines and enforces the environmental aspects of monitoring, data collection and processing
- Defines harmful water pollution and the permissible level of pollutants
- Develops regulations for water protection and enforces them
- Defines environment related tasks in preventing and reducing damage from accidental water pollution
- Encourages environmentally safe water management.

In the scope of waste management, the Minister:

- Defines environmental and professional requirements for the treatment and disposal of industrial and municipal waste
- Directs protective activities against harmful effects of waste
- Takes part in preventing and reducing waste production, in organizing recycling.

In the scope of noise protection, the Minister:

- Defines rules for setting standards of protection against noise and vibration
- Takes part in defining standards.

The new Environmental Act specifies the task of developing environmental strategy as an elaboration of the National Environmental Programme based on an analysis of environmental trends. It also adds to the set of general responsibilities the requirement of actively supporting development and implementation of environmental curriculum in education.

There are, however, other ministers with some authorities in the field of environmental protection.

The *Minister for Public Health* also has important tasks in the field of environmental protection:

- Defining the tasks related to prevention and reduction of adverse health impact of environmental pollution
- Defining ambient environmental standards
- Directing the activities of the National Public Health and Medical Service.

National Public Health and Medical Service through its regional network carries out monitoring and some enforcement activities in the field of environmental quality and waste disposal.

The *Minister for Transport, Telecommunication and Water Management* has the overall responsibility for water management and enforcement role in transport related air quality protection. The *Minister for Industry and Trade* has the co-ordinating role in hazardous waste management and the supervising role over the network of hazardous waste treatment plants. In the last four years, the content of this supervisory role is in transition due to uncertainties brought about by massive privatization in the definition of sector tasks.

The *Minister for Internal Affairs* has environment related tasks through the relationship with municipalities in:

- Directing municipalities' activities in the field of municipal solid and liquid waste treatment;
- Preparing suggestions for yearly direct government subsidies in those areas.

Municipalities have the authority to set stricter environmental standards and requirements than the ones stated in national legislation. They must ensure enforcement of national standards and policy goals. They must develop local environmental programmes and analyse local environmental trends. They must report their findings to the public at least once a year. The municipalities are in charge of developing an emergency plan for smog alert and restrict the fixed air pollution sources accordingly. Municipalities are also responsible for management of municipal solid and liquid waste. Therefore they supervise and often own sewage treatment plants and waste treatment facilities.

The National Council of Environmental Protection is a newly created institution. Its role, as defined in the Environmental Act, is to create a regular forum for public participation and scientific, social and professional standards for environmental protection. The Council reviews and takes a stand on basic principles of various environmental programmes, environment related legislation and decisions. The Council then submits its position to the Government. Since this National Council is in the process of establishment, there is no experience with its scope of operation and the actual impact of its activity.

The main body for enforcement is the Main Inspectorate for Environmental Protection which is the second level legal authority in environmental matters. The local bodies for enforcement are the Regional Inspectorates which are the first level of legal authority. They are in charge of monitoring, collecting information, setting and collecting certain fines. Nature conservation has been separated from the Inspectorates, and three nature conservation authorities and five national park authorities have been formed. In connection with the new Environmental Act, a revised set of tasks and responsibilities for these regional enforcement bodies should be developed.

In the new organizational setting some tensions originate from the traditional institutional lines of environmental protection. The National Institute for Public Health was established in the mid-1920s with the task of identifying and preventing health hazards. A related network of public health inspectorates was established. Since pollution was acknowledged as an important factor in causing health hazards, these organizations play a significant role in environmental protection. In connection with public health issues, water protection also appeared early as a prominent element in the operations of water management. This history deserves notice because it shaped the ambitions of some organizations in the environmental administration. There is still a lingering view that the Ministry of Environment and Regional Policy and its organizations can assume full responsibility (control) for environmental protection only if they organizationally incorporate earlier environmental related tasks. It makes co-operation with other ministries sometimes difficult and can hinder integration of environmental concerns into sectorial policies.

Another critical issue is the separation of water management and water protection. Clear demarcation of responsibilities and the realm of operation has not taken place. The executive orders derived from the decree and related to the tasks and responsibilities of the two ministries overlap at many points.

Incomplete reorganization of Inspectorates creates enforcement problems. The area of Inspectorates corresponds to the water catchment area, not to the border of the counties. Therefore, it is difficult to fit local environmental protection into the general public administration structure. Another problem is the lack of 'institutional memory' at the Inspectorates including information on the history of several pollution cases. As part of the public sector reform, changes in the structure and area of regional environmental inspectorates are planned but a new design has not yet been elaborated.

Environmental policy

The environmental legacy of the socialist, centrally planned system is usually described as disastrous or devastating. The use of dramatic terms and shocking pictures dominated discussions about the environmental situation. Estimates in the range of billions of dollars appeared as the cost of the necessary remedial actions and clean-up measures. Such a tone could be certainly justified, but it overshadowed two important features of successful environmental policy for the transition period. One is the need to protect those areas which were left relatively untouched by inefficient socialist industrialization. The other is the possibility to utilize market forces for environmental protection and save valuable resources, not just emphasizing the need to spend a lot of money on it.

What is the 'free environmental gain' we can expect from developing a market system? Two features of industrial development under centrally planned resource allocation are worth mention here. One is the concentration on heavy industry with the added feature of autocracy and a predetermined economic role inside the CMEA countries with no particular regard to an individual country's natural resource endowments. The other is inefficient use of resources, as energy or any other input material, in the production process. The apparent signs of inefficiencies in resource utilization have prompted research and modelling work to solve this issue.

There are two potential 'free' gains for environment from the switch to a market economy in Hungary, as well. The country's resource endowments are not well suited for heavy industry, therefore, market driven restructuring clearly increases the role of service sectors which are generally less polluting. Meantime, the rigour of market price mechanism can be expected to induce efficiency gains in general resource use including energy use and natural resource utilization.

These two elements of transition certainly can bring substantial improvements, but there is need for some safeguarding as well. The logic of price mechanism implies that profit maximizing or cost minimizing behaviour shifts input use towards cheap resources. We find, even in developed market economies, that cheap resources often include natural and environmental resources. Their low prices, however, can reflect the deviation of market prices from the social valuation of those resources. Meantime, one of the most important features of the transition, namely privatization, introduces temporary uncertainties in the property titles. Unclear or transitory property right assignments encourage over-exploitation of natural resources, such as excessive logging in some forest areas or short-sighted cultivation practices in the formerly state owned or co-operative land. These possibilities call for specially designed environmental policy measures.

A study has been carried out to examine the environmental impact of economic restructuring in Hungary (Abel et al., 1993). It has found that with respect to sectorial composition, potentially polluting industrial subsectors, such as paper, or chemical industry, do have comparative advantages – both short and long term. Concerning the micro level processes, the most obvious ways of improving efficient use of resources already have been exploited. Further steps in resource savings now often require radically different thinking and approaches and/or new investments.

The environmental protection challenge encouraged the Parliament of the Hungarian Republic to ask the Government to develop the National Environmental and Nature Conservation Policy Concept (NKTK). The Concept was prepared and accepted by Parliament. It serves as a basis for the development of the National Environment Protection Programme for six years' duration as legislated in the Protection of the Environment Act.

The document lists principles for environmental protection policies rather than describing actions to be taken. It lists priority environmental problems, but even the problems needing immediate attention are very long and do not indicate any focus. Tasks of the short-term environmental protection priorities are:

1. Map activities (industrial plants) that pose direct and severe hazards to human health and environmental values in large areas.
2. Clean-up environmental pollution and damages accumulated during the last decades, following specific programs. Regular financial funding must be made available for the eradication of pollution.
3. Halt the deterioration of the environment in regions which are subjected to increased environmental load and cumulative pollution effects. Improve environmental protection in harmony with the implementation of structural transformation, regional development and social programs.
4. Establish domestic systems of environmental rating, labelling and auditing following recommendations made by the Office of International Standards and by the European Union. Assist the development of the handling of the environment and companies' environmental management systems.
5. Protect biological diversity through management of the most important nature preservation areas as properties of the State.
6. Improve the quality of air in polluted areas, and reduce air pollution in settlements. Special emphasis must be made to reduce air polluting impacts of traffic. The Air Action Program for 1994–98 which was approved by the Government defined specific tasks. Amongst those, the following are especially important:
   - develop concrete action plans in the specified regions;
   - adapt EC rules of law regulating road vehicle emissions as a precondition of licensing roadworthiness of new vehicles;
   - reduce vehicle emissions of the existing fleet through retrospective technical actions and intensified supervision of the operational standards;
   - moderate air pollution in large cities by improving the standard of services offered by public transport and through traffic control measures;
   - continue the development of air pollution monitoring networks for the measuring of background pollution within the settlements. Develop an out-of-settlement monitoring network in regions with sensitive ecological systems;
   - fulfil international agreement obligations.

7. In order to reduce the hazardous effects of traffic noise, the implementation of passive acoustic protection must be started in areas with the most severe load.
8. Develop a Government program for protection of drinking water.
    −   Improve the water balance in regions endangered by the permanent drop in water level (e.g. the area between the Danube-Tisza, Trans-Danube Central Hills, etc.),
    −   Reduce the load of large amounts of polluting substances originating from the infrastructural gap between piped water supply and sewerage collection, primarily by significant development of sewage systems and waste water treatment.
    −   Motivate water-saving solutions through technical development and information. Provide central grants and other support (e.g. incentive-development support grants).
9. Agricultural soil is a resource of limited availability, therefore, efforts must be made to protect its quality and quantity.
    −   In land use regulation, primarily preserve the land's prolific role.
    −   Halt increasing soil acidity through selection of the appropriate chemical fertilisers and by strict regulations on the use of chemical fertilisers.
    −   Application of the appropriate agricultural technologies and returning organic matter into the soil are necessary for the preservation of humus and fertility of the top-soil.
    −   Prevent or moderate soil losses caused by water and wind erosion through protective measures for cultivation of local soil properties and use of land.
    −   Encourage soil regeneration in areas removed from agricultural use.

(from the National Environmental and Nature Conservation Policy Concept)

Changes in environmental legislation

In the field of legislation, there have been fewer advances than expected. The main problem has been the long period during which the comprehensive environmental Act was prepared and negotiated. The delay had some institutional reasons, but it was most likely due to somewhat unexpected difficulties in redefining the scope, means and institutions of environmental protection for the emerging democracy and market economy.

This delay overshadowed all other regulatory legislation. In the first five years of transition, the only Acts passed were the Act on the Gasoline Product Charge and the Act on the Central Environmental Protection Fund. In the field of technical regulation, basically everything was slowed down to wait for the environmental framework Act. There have been amendments to old legislation related to hazardous waste management, water quality control, and air quality protection, but they did not bring the necessary changes. After three years of waiting for the comprehensive environmental Act, a Ministerial Order in 1993 was passed on Environmental Impact Assessments (EIAs).

In May 1995 the Act on the General Rules of Environmental Protection was passed and came into force toward the end of 1995. It is a framework law that defines fundamental rules and general requirements for environmental protection. The Act prohibits polluting air, water and soil. Polluting sources shall not emit any pollution, at least no more than the permissible level.

Reflecting the changing nature of the economic and social institutions, particularly private property and democratic institutions, the new Act stipulates the need for co-operation among state authorities, local governments, private individuals, their organizations, enterprises, and trade organizations. The language, however, is rather authoritative (everybody has to co-operate). The Act is somewhat weak on presenting procedures or institutions for such co-operation in which affected parties would be involved through their interests and not just by the hardly enforceable words of the Act. Important new elements of the Act are:

- Environmental information has to be made accessible to the public.
- It is the responsibility of state authorities and local governments to collect information on the state of environment and its impacts on human health.
- The citizens' right to environmental education and knowledge is emphasized. It is the responsibility of the State at the national and local level to establish conditions for exercising that right.
- The requirement for a public hearing if a detailed environmental impact assessment has to be done is introduced.
- The requirement to elaborate on the details for effective public participation.

In the Act very little is stated about implementation and, particularly, enforcement of the general (and fairly strict) rules and aims. There are few additions to the previous, not too effective enforcement toolkit of environmental protection. Limiting or closing down polluting activities and levying non-compliance fees (fines) still appear as the backbone of enforcement. The possibility of requiring the establishment of an escrow account for covering the cost of future environmental remedial measures and/or requiring environmental liability insurance is a new element.

Several technical regulations related to different standards and target values for the state of each element in the natural environment still must be developed and legislated. These detailed rules will ultimately shape the scope and possible options for environmental protection. Even though the new Environmental Act shows clear signs that several aspects of environmental protection have been adjusted to the new social and economic situation, in many respects a 'command and control' type approach is reiterated. In spite of the extended set of possible economic instruments introduced by the Act, the scope for using these instruments as cost-effective incentives seems rather limited. Since the Act calls for emission standards as important pillars in environmental regulation and strictly prohibits emission above those standards, there is not much possibility left for the flexibility which is required for the efficient operation of an economic instrument. Economic incentives are cost-effective instruments if polluters have a real choice on the level of pollution allowed and are not bounded by strict emission standards and policy-makers concerned with ambient environmental quality standards. Presently, it looks more as if environmental policy will be based on strict emission limits which come from some form of 'best available technology' and emission charges. Product charges will have the sole role of raising revenue for an earmarked environmental protection fund.

Economic instruments

It is often suggested that massive economic and institutional restructuring in Central and Eastern Europe offers the chance to avoid costly mistakes made in environmental policy in developed market economies. Institutionalized 'command and control' measures have contributed to establish the view that good environmental policy must tell each polluter what to do and enforce it. On the flip side, environmental protection has been inevitably perceived as a request for costly investment which does nothing but constrain economic growth.

Experiences of developed countries teach us that prevention is cheaper than 'end of the pipe' solutions; that economic self-interests are often more efficient incentives than strict commands. To move environmental administration and regulation towards better integration into economic policy and regulation, more extended use of economic instruments would seem, therefore, efficient. However, to act according to these principles: (i) requires good foresight into the future economic structure, i.e. short- and long-term predictions; and (ii) assumes a substantial co-ordinating role in different sectors.

Both these requirements are hard to fulfil in a period of economic and social restructuring. One consequence is that due to uncertainties associated with the two conditions above there is a strong tendency to move environmental administration into a 'command and control' type pattern. This certainly can be observed in the new Environmental Act.

In Hungary there is no example of a direct emission tax. However, product charges have recently figured prominently, and the fines aiming at enforcing different environmental norms and laws are in place. Extraction charges, which can be viewed as environmental tax, also must be paid for exploiting certain environmental resources (see Table 4.11). There is also a precedent in the tax system for differentiation on environmental grounds in the case of both turnover and income type taxes.

Product Charges

By the beginning of 1996, a set of new product charges came into effect. Tyres, refrigerators, refrigerants, batteries, packaging materials and gasoline

Table 4.11   Revenue from environmental charges, fines and fees in Hungary (million forints)

	1993	1994	1995 expected	1996 predicted
Revenues from product charges				
Transport fuel	1 460	2 502	3 400	4 800
Tyres	–	–	150	900
Refrigerators/refrigerants	–	–	80	480
Packaging	–	–	–	4 000
Batteries	–	–	–	700
Total:	1 460	2 502	3 630	10 880
Revenues from environmental fines to CEPF				
Air pollution fine	394	470	350	350
Wastewater fine	156	152	200	200
Hazardous waste fine	44	29	40	40
Noise and vibration fine	25	11	20	20
Total:	620	662	610	610
Other revenues from environmental fines				
Municipalities' revenues from environm. fines	...	...	519	600
Sewage fine[a]	...	200	200	200
Total:[a]	...	200	719	800
Revenues from natural resource extraction fees				
Mining extraction fee[b]		14 249	15 000	15 000
Water extraction fee		2 900	2 950	3 300
Forest extraction fee		1 823	...	2 310
Land protection fee		296	...	370
Total:		19 268	17 950	20 980

Source: Lehoczki and Morris (1995).

Notes: [a]Estimates; [b]5% of the revenue is separated for CEPF.

are subject to environmental charge. The general structure of these charges is similar except that for packaging materials. The charges must be paid on the subject products at the first point of sale or use of domestically produced products or else, in the case of import products, paid by the importer. Among the subject products, those receiving an environmentally-friendly or an eco-label are given a 50% allowance.

The product charges are to be paid to the Central Environmental Protection Fund (CEPF). In the case of domestically produced products, the administrator of the CEPF is to check the execution of tax assessments and payments, while in the case of imported products, the Customs Office is to fix and collect the product charge.

The product charge revenues are spent through the CEPF. In accordance with the CEPF's spending regulations, a part of the income can be used for financing or subsidizing environmental protection projects. However, at least half of the incomes are earmarked even within the CEPF. Thus, the earmarked proportion of sources coming from product charges should be spent on subsidizing solutions to the environmental problem caused by the subject product. Permanent subsidies are a new form of subsidizing the CEPF can use under the new product charge law which allows a continuous contribution to current (operational) costs.

The *Environmental Product Charge on Transport Fuel* was the first economic instrument to aim at taxing the consumption of a product for environmental protection purposes. The law became effective in May 1992. Transport fuels cause air pollution when consumed. Direct emission taxation is infeasible because of the numerous and mobile sources (vehicles) and cost of measurement. Therefore, a tax imposed on transport fuel as an input to pollution production can be viewed as a feasible, close substitute to a direct emission tax because of its low administrative cost and high correlation with emissions.

Nevertheless, in practice, as made clear in the introduction of the law, the product charge provides financial sources necessary for projects that reduce environmental damage caused by transport, although it also has an incentive effect on pollution-producing activities by reducing the demand for motor fuel. The amount of the charge has gradually increased from the initial 0.5 HUF l^{-1} to 2 HUF l^{-1} (US\$ 0.006 to US\$ 0.016). Taking into consideration that even this increased tariff is only 1.8–2% of the consumer's price, while the whole tax content and the road fund contribution together reach 70–75%, its independent effect on motor fuels obviously must be modest.

The *Tyre Product Charge* taxes tyres of automobiles, agricultural tractors, construction equipment, trailers, and aeroplanes, whether they enter into circulation independently or as part of the vehicle. The size of the charge is 30 HUF kg^{-1} (US\$0.2) for new tyres; for imported used tyres the price is 150 HUF kg^{-1} (US\$1). The amount of the charge for used tyres imported for re-trading is determined by a separate law. In the first place, the aim of the product charge is to raise revenues for financing the environmentally-friendly disposal of used tyres. The size of the charge was determined on the basis of this aim. Few attempts were made to quantify the environmental damage connected to the usage of tyres and to fix the amount of the charge accordingly.

Product Charges on Refrigerators and Refrigerants aim to solve two separate environmental problems. One problem is the inadequate disposal of used refrigerators and the increasing quantity of solid waste. The other one is the air pollution caused by ozone-depleting refrigerants. These polluting substances escape into the air during repair or disposal of refrigerators. The object of the product charge is to contribute to the proper recovery, re-use, or safe disposal of ozone-depleting refrigerators and to recycling of certain parts of refrigerators. The product charge supplements the existing direct regulation on the import, production and use of ozone depleting substances. The main purpose of the product charge is not in reducing the use of ozone-depleting substances since the law forbids this, but rather in raising revenues for collecting and treating ozone-depleting substances already in use for refrigeration.

The *Environmental Product Charge on Packaging* is intended to encourage the re-use or recycling of packaging. The charge rate ranges between 2 and 10 HUF kg^{-1} (US$ 0.016 to US$ 0.08) depending on the material. The party obliged to pay the product charge is exempt from paying when a target proportion – fixed in a separate law – of packaging material is collected, repurchased and re-used or else reprocessed or combusted for energy generation. The product charge obligation, and thus the exemption, can be passed to a treatment co-ordinating organization.

The likely aim of the *Environmental Product Charge on Batteries* is to raise revenues for projects that solve battery-associated environmental problems. The amount of the charge is 38 HUF kg^{-1} (US$ 0.03).

Environmental Fines

This economic instrument for environmental regulation is the longest and most widely used in Hungary. The level of the fines does not encourage compliance. Its amount is extremely low both absolutely and when related to the fact that in many cases fines should result in a 50–60% reduction in the emissions of polluting agents. Their revenue-raising function, however, continues. Levying of fines is usually based upon self-reported emissions. The basis is the magnitude of deviation from individual emission standards and depends on regional factors. Monitoring self-reporting and measuring the actual emissions is rarely possible for such a large number of pollutants. Furthermore, with the change in Hungary's economic structure, the number of minor polluters increased. To evade the fines became easy. This lesson must be taken into consideration when designing direct emission charges.

The *Sewage Fine* is to be paid by the legal entity organization or venture based on the amount of polluting sewage discharged into public waterways. The *Waste Water Fine* is paid by the legal entity organization or venture that is emitting contaminating or damaging polluting sewage into surface waters. The *Air Pollution Fine* is paid by those polluting operators whose facilities are located at a fixed place and who exceed the emission limit determined for them. This fine does not involve households. The fine can be imposed on cars used in traffic in cases where they exceed the limit emissions determined by regulations. The *Noise and Vibration Fine* can be imposed on operators or contractors during their building or reconstruction work if they exceed the determined noise emission limit or vibration burden limit. The *Hazardous Waste Fine* can be imposed on the producer – except health institutions and private persons – if the regulations referring to reporting, collection and pre-treatment, transportation and temporary storage or neutralization of dangerous wastes are broken.

There are special fines related to regulations controlling and phasing out the use of *ozone-depleting substances*. The fine should be paid by the user for any unpermitted use of ODS. The *Fine on the Import of Hazardous Wastes* makes possible the imposition of fines for not obtaining permits related to the import of dangerous materials or activity different from that permitted. The *Land Protection Fine* aims to protect agricultural areas. It fines those who, through their own fault, do not use agricultural land, do not maintain its productivity or pollute the soil with damaging substances. The *Air Pollution Fine for Mobile Sources* covers those polluting substances for which ambient air quality norms have been established (such as CO, NO_x). The fine is levied and fixed by transport authorities. Basically, it is not enforced.

Environmental Aspects of Tax Differentiation

Preferences in the tax system are the most widely known tax-type instruments for environmental protection and, therefore, they are frequently recommended by environmental policy-makers. The concrete form of differentiation and allowances in many cases change annually. This can easily undermine the incentive impacts, that is, the direct environmental effectiveness of these instruments.

In the 1995 Corporate Income Tax Law, there are some important, environmentally-motivated tax allowances. One is the accelerated amortization of environment protecting machines, equipment and apparatus. Another important allowance is increasing the financial reserve to cover possible environmental obligations determined in laws on mining, electricity production, transport, and service.

If environmental services such as solid waste management and wastewater treatment are provided by 'companies serving public interests', the company is exempt from the corporate income tax. Corporate income tax allowances connected to these or similar activities were established in 1990, therefore, they can also be considered to have been incentives for some time.

The possibility of obtaining substantial tax allowance related to environmental investments occurred twice. In 1991, 20% of the environmental investment cost could be retained from the calculated corporate income tax. Although it was only possible to gain access to this allowance with investments starting in 1991, a significant impact occurred in the 1992 tax payments. The other investment allowance was offered in 1994 to companies making large environmental investments if, as a result of the investments, more than half of their income came from the sale of environmentally-friendly products. The size of the tax allowance reaches 100% in the first year and 60% in the second year. The allowance's environmental incentive impact was weakened by the fact that the allowance also could have been obtained by performing different – non-environmental – investments. There were few examples where enterprises requested tax allowance by representing their investment as job creating or serving export expansion rather than trying to fulfil the conditions for producing environmentally-friendly products.

Most environmentally-motivated consumption tax differentiation is connected to transport. Its basic characteristic is that it encourages public transport. In the case of cars with catalytic converters whose cylinders displace up to 1600 cc, and electronic cars, a 0% tax rate as opposed to the general 12% tax rate should be applied. Cars with catalytic converters above 1600 cc have a 10% tax rate rather than the average 22%. The consumption tax on unleaded gasoline is lower than that of leaded gasoline by more than 6 HUF (US$ 0.06). In 1992, vehicle excise tax was reduced by 50% for vehicles with catalytic converters.

Some environmental goods and services are classified under a preferential VAT rate (12% as opposed to the general 25%). Included are activities such as performing EIAs and environmental audits, and products, such as solar energy collectors and catalytic converters.

Privatization, bankruptcy procedures and the environment

The privatization Act as first passed and applied hardly took any notice of environmental liabilities. Environmental liability issues were perceived as part of the general liability clauses in the privatization contracts. The foreign investors were, however, more aware of the potential legal and financial consequences of discovering some contaminated sites or hazardous waste dumps. They requested and received full protection in case of such events. In one case, the privatization agency gave full financial coverage for cleaning up contamination discovered after the privatization deal was signed. When contaminated sites were discovered, the new owner would clean them up to the highest standards and present the bill. Such expensive lessons proved that it is not wise to ignore environmental liabilities. Therefore, the privatization agency is more willing to get engaged in regular discussions on the matter.

At later stages of privatization legislation, environment is mentioned in the context of the business plan that may be required of, and prepared by, a prospective new owner. At that time, the new owner may be required to include the assessment of environmental liabilities associated with the property together with a future plan for handling the problem.

The separation of accumulated past pollution from the 'ongoing' pollution is a difficult problem. Responsibility for past pollution is usually addressed in some form of environmental or general liability legislation. In the period of transition, it became a particularly pressing environmental policy problem due to the large volume of ownership changes (privatization) and bankruptcy cases of inefficient enterprises. The number of bankruptcy procedures was below 100 in 1992 but grew above 1000 by 1994. Regional Environmental Inspectorates are the ones who should guard environmental interest in that process. Environmental claims in the bankruptcy processes increased from US$0.1 million to US$1.5 million (KTM, 1995) during the same period. These claims consisted mostly of unpaid environmental fines. Lately, they have included subsidies from the Environmental Fund which need to be paid back.

Contaminated sites and accumulated hazardous waste in bankrupt enterprises are becoming a problem. The cost of cleaning up these sites or properly treating hazardous waste is estimated at several billion HUF. There has not been, however, a comprehensive approach or policy developed towards surveying the problem and financing solutions.

Financing environmental protection

Figures for environmental spending usually cover investment costs only. Attempts have been made, particularly at the encouragement of OECD experts, to estimate operation and maintenance costs as well. Results are not extended and convincing so far. Attention is also focused on investment spending which approximates the size of environmental protection activities.

The level of environmental investment at current prices and its allocation across environmental media is shown in Tables 4.12 and 4.13. Data show fairly stable levels of investment at current prices between 1988 and 1991 but indicate declining real value. We can observe a large variation in the real value of environmental investments. It seems that the real triggering force for increased investment is launching new subsidy programmes.

More than half of the environmental investments deal with water protection. Air quality protection and waste-related investments both are around 12–18%. Hardly any tendency can be traced from the chain of increase and decrease. The share of public sector in investments is very high, around 35–40%, and is almost entirely sewage related. The power sector's environmental investments are concentrated in air pollution control.

The sources of finances are direct budgetary allocation. The most important among them is the directed and targeted subsidy programme for municipalities. In the years when the programme covers environmental targets such as sewage systems and solid waste management, investment spending increases substantially.

An important financing source is the Central Environmental Protection Fund. The CEPF is a general, broad based national environmental fund with the aim of assisting all areas of environmental protection and nature conservation. As stated in its charter, it is established as an extra-budgetary state fund with the aim of encouraging the development of an environmentally sustainable economic structure, preventing environmental hazards, eliminating or at least reducing environmental damage from past pollution, maintaining and developing protected nature conservation areas, promoting conditions

Table 4.12 Environmental protection investment indicators for Hungary, 1988–94

	1988	1989	1990	1991	1992	1993	1994
Environmental investments, million HUF	12 088	12 827	12 336	11 708	20 838	19 661	37 777
Environmental investments, million US$	240	217	195	157	264	214	359
Environmental investm. as a % of GDP	0.84	0.75	0.59	0.51	0.75	0.59	0.95
Environm. investm. as a % of total investm.	4.16	3.78	3.46	2.38	3.75	3.08	4.48
Environmental investments/capita (US$)	22.6	20.5	18.8	15.1	25.6	20.7	35.0
Exchange rate HUF/US$	50.4	59.1	63.2	74.7	79.0	92.0	105.1

Sources: Koloszár and Lehoczki (1993); KSH (1995).

Table 4.13 Share (%) of different types of environmental investments in Hungary

	1987	1988	1989	1990	1991	1992	1993	1994
Protection of cropland	16	15	14	14	4	4	4	5
Water protection	56	50	53	52	45	60	60	61
Air quality protection	10	13	10	12	32	10	10	6
Living world protection	0	0	0	0	1	2	2	0
Nature conservation	0	0	0	1	3	4	4	9
Settlement environm. prot.	3	3	3	5	6	4	4	3
Waste related	13	15	18	14	7	15	15	14
Noise protection	1	3	2	2	1	1	1	1
TOTAL	100	100	100	100	100	100	100	100

Sources: Koloszár and Lehoczki (1993); KSH (1995).

Note: From 1992 on, the data cover only organizations with more than 50 employees. Therefore, comparison between data before/after 1992 should be made with care.

necessary to implement efficient environmental protection measures, as well as fostering environmental awareness of Hungarian society.

The CEPF is not an independent financial institution. It operates under the auspices of the Ministry of Environment and Regional Policy (MERP). The Minister of Environment and Regional Policy is responsible and accountable for the Fund to the Parliament. Table 4.14 provides information on the size of different sources to the Fund.

In 1991, 70% of wastewater fine revenue became designated as a source for the CEPF to be spent on water quality protection measures and subsidies. This was the first strict earmarking of a revenue source inside the CEPF, even though it was always an informal requirement to maintain the similar proportion of subsidies across environmental media as the proportion of fines levied across different media.

In 1992, an important new factor was the addition of the environmental product charge on transport fuel as a revenue source. The charge on transport fuel became the first revenue introduced with the declared objective of raising money for a particular environmental programme, namely addressing transport related pollution problems. This shifted emphasis in the Fund towards organization and methods for disbursement.

Different spending categories called 'windows' are defined in the legislation. One type of spending is assistance for private and public sector activities that aim at direct improvement in environment in any of the targeted areas. Assistance provided through this window can be grant, soft loan or loan guarantee.

Other types of spending are ones related to important state tasks in the field of environmental protection which must be financed somewhat independently from fluctuations in the environment section of the state budget. Such state tasks include monitoring and raising public awareness.

Foreign assistance provides only a small fraction of resources for environmental protection. The major source has been the PHARE Assistance Programme (see Table 4.15). The other important source is the Japanese soft loan provided for environmental infrastructural development in one of the old industrial regions near Lake Balaton.

Private sector participation in environmental protection

The number of enterprises sharply increased in the first three years of transition reaching 99 044 by 1994. The growth was 52% in 1992 and stabilized at around an annual increase of 25%. Meantime, the number of enterprises employing more than 300 persons declined from 2617 to 1340 between 1989 and 1994 (KSH, 1995). The number of enterprises which can be considered as polluters has grown at least at the average growth rate, or slightly more, and the number of traditional big polluting enterprises decreased probably by half.

The environmental impacts of the increasing role of multinational corporations is a widely debated topic. In several cases, companies have built on their

Table 4.14 Revenues to the Central Environmental Protection Fund (CEPF), 1988–96 (million HUF)

	1988	1989	1990	1991	1992	1993	1994	1995	1996
Fines	495	387	444	1121	552	575	663	610	610
Transport fuel product charge	...	...	...	...	879	1 460	2 502	3 400	4 800
Other product charges					...	...	...	230	6 080
Part of transit traffic forgone by the state	...	...	...	...	204	663	...	0	0
Direct budget allocation	...	...	61	255	0	0	0	0	0
Foreign assistance: PHARE contribution	...	...	...	...	...	...	314	470	0
Part of mining fee	...	...	...	...	...	...	2 496	660	840
Loan repayment and recalled support	...	...	...	.	...	...	105	135	355
Interest earned in short term investment	...	...	...	...	...	...	901	1 500	0
Loan take by the Fund	...	...	...	...	...	...	338	0	...
TOTAL	495	387	505	1 376	1 635	2 698	7 317	7 005	12 685

Source: Government Budget (1993, 1994, 1995, 1996). Note: see Table 4.12 (p. 160) for exchange rates HUF/US$

Table 4.15 Use of Phare environment sources in Hungary by sectors and phases (million ECUs)

	I. phase	II. phase	III. phase	V. phase	Total
Institutional development, Education, Public awareness	1.08	2.07	0.57	3.50	10.22
Air pollution, Transport, Noise control	6.24	4.38	0.63	3.00	11.25
Water quality control	6.93	–	0.60	–	7.53
Waste management	1.91	2.16	0.40	–	4.47
Nature conservation	2.63	1.00	0.40	–	4.03
Energy saving	5.12	–	0.40	–	5.52
Fund for investments, CEPF	–	–	6.30	7.50	13.80
Other	2.54	0.80	0.70	0.50	4.54
TOTAL	26.45	10.41	10.00	14.50	61.36

Source: Donáth (1995).

experiences in developed countries and their practice is ahead of environmental regulation requirements. They often respond to future planned regulations which is rarely the case with the Hungarian firms. However, these multinational corporations are also ready to capitalize on the weaknesses of the environmental regulations. There have been cases when products or technologies appeared in their plants which were already environmentally unacceptable in their own countries.

The number of enterprises in the environmental 'service' sector is difficult to determine because the new statistical classification system does not have an 'environmental service' category. In 1992, 342 enterprises indicated wastewater or waste treatment as their main activities in their corporate tax statements, and their number is growing.

In 1994, an REC survey was conducted on the Hungarian environmental business sector (Dzuray, 1995). It reported about 400 companies that provided environment related services or products. Their activities were as follows:

- Technical services, particularly engineering and design for air protection
- Environmental product related activities, particularly products for air protection
- Laboratory activities.

The share from reported companies' revenues is the largest for water related activities, followed by air related activities and waste related activities.

Public perception and participation

There have been several smaller surveys which attempt to monitor social environmental awareness. In the mid-1980s, there was a big research project which aimed at assessing the possibilities for realizing environmental interests. The results showed that some recognition of environmental problems was already present but only at a fairly high level of generality. A representative survey in 1988 indicated that 79% of the respondents preferred a society which saves natural values for the future generation and 62% would put more emphasis on protection of the environment than on production (Lehoczki and Szirmai, 1988). Those clearly affected by local pollution showed more consciousness, for example mothers of young children in heavily polluted industrial cities. People, however, indicated that only high level officials and leaders could act to initiate changes (Kulcsár, 1991).

Recent studies indicate that several specific environmental problems are widely known, such as global warming, acid rain, and air pollution in the big cities. People, however, perceive environmental degradation as a problem they can do nothing about. They do not consider how their individual behaviour contributes to pollution. The usual justification is that they cannot afford to contemplate environmentally-friendly choices. A survey made by the Gallup

Institute in 1994 indicated that sensitivity for environmental issues had increased in the period of transition but was not as developed as in some market economies. This was revealed in several cases as well. For example, the participation rate in recycling programmes was fairly low and demand for environmentally safer products was sluggish.

With respect to media coverage and public awareness, two tendencies can be observed. Building up civil society increases open recognition of local interests in environmental problems. This results in bringing more environmental issues into the light. The second tendency, however, works for covering the fact or the impacts of pollution and it is due to economic interests.

Therefore, the structure and flow of information reflects how environmental policy is developed and administered. The fact that sweeping political changes are taking place in Hungary suggests that we can expect alteration in that process. As the first sign, a number of environmentalist groups are asking for more opportunities to shape environmental policy; non-governmental organizations are putting forward the claim of advising and/or checking up on environmental policy-makers. However, all these requests will not bring about sound changes in the efficiency of protection measures unless these groups and organizations gain access to appropriate and reliable data. Since the prevailing system of collecting, processing and storing information related to air pollution is still in line with centralized decision-making, it is an urgent task to restructure it. However, we cannot agree with those experts who believe only in developing a new, independent information system for non-governmental use. Discarding 'official' information rather then rebuilding it seems a waste of public money.

We can illustrate development in public participation processes through the case of designing and negotiating the environmental product charge legislation. Based on initial experiences with the Product Charge on Transport Fuel, the Ministry of Environment and Regional Policy in 1993 initiated legislation on product charges for tyres, packaging material, refrigerators and refrigerants.

The most extended among these proposed charges has been the one on packaging materials. The proposal was attacked by several trade organizations and environmental NGOs at the same time. Some criticized the Ministry for putting too large a financial burden on producers and consumers, some for not being comprehensive, and some for applying differential rates. The Parliament's Environmental Committee, however, reviewed the legislation and requested revisions to expand the scope for every packaging material. The revised version was negotiated at the end of 1993. The negotiations ended with an agreement between the Environment Ministry and several trade organizations representing the most affected companies, several of which were multinational corporations.

In 1994, many different revised versions of the proposed legislation were prepared and discussed with reference to the signed agreement. These were often criticized by NGO representatives for reflecting too much the views and will of the trade organizations. By the beginning of 1995, the newly elected leadership of the Ministry decided to strike a compromise with the trade organizations. Meantime those organizations prepared themselves to be 'constructive' because they did not want to be perceived (mostly the multinational companies) as not caring for the environment. New rounds of negotiations resulted in a new agreement in which the trade organizations agreed to the extended scope of the charge. The Ministry agreed to an extended set of possibilities for waiving or reimbursing charge payments and reconsidering the product charge in the context of comprehensive solid waste management legislation. This was largely criticized by NGOs, but they allied with the trade organizations to get the revenue earmarked for financing recycling and re-use of packaging material. The legislation calls for establishing a broad based, national organization for overseeing the collection and use of revenue.

In the case of the other product charges, negotiations were far less extensive due to their much more limited impact. One interesting element in the refrigerant product charge is that the big refrigerator producers and importers supported the introduction of the charge, due to the existing strict regulation on phasing out ozone-depleting substances. This is expected to be enforced mostly on the well-known, big companies but not too much on the small firms. The big companies felt that they could minimize the impact of that disadvantage if a product charge were levied, as well as collected from small firms. At the same time there was a strong lobbying effort to earmark the revenue to subsidize strategies for compliance with the ODS legislation.

The main lesson from the preparation and negotiations of product charges is that the Government must face an increased number of new forms for articulating interest. The Government has to be prepared to formulate its proposals in a fairly complicated process that is sometimes dominated by strong economic pressure groups or negotiations which are not based on strong professional assessments.

4.4 CONCLUSIONS: CHALLENGES FOR ENVIRONMENTAL PROTECTION

Overall assessment of the changes in environmental quality and use of natural resources shows a positive picture over the five years of transition. Emissions, pollution concentrations and energy use generally declined. There are, however, strong indications that these encouraging trends are mostly due to economic recession rather than the results of environmental policy measures. Environmental policy-makers may get reasonably high scores in such areas as broadening the financial base for environmental protection, signing international agreements and opening up many institutional ways for public participation. Nevertheless, the real strength of these achievements is still to be proven. Meantime, responses to major challenges are delayed or inadequately addressed.

One of the major challenges is the better integration of environmental policies into economic and sectorial policies. The concept of 'win–win' environmental policy measures which means actions profitable in both an environmental and an economic sense is widely propagated in the Environment for Europe process. (It is a centrepiece for the Environmental Action Programme for Central and Eastern Europe, presented at the Environment Ministers' Conference in Lucerne in 1993.) The method of developing an environmental strategy took a different direction. Priority actions have been based on the scientifically desirable state of environment after the year 2000 with the wish that the economic development should adjust to those environmental priorities. It has been ignored how difficult is to induce radical environmental protection measures when the main concern of the industrial plants is survival. Therefore, even if the starting point for environmental policy is that environmental protection should not be subordinated to economic interests, it can be realistic only if economic constraints are taken into account. The details of this lesson are still to be learnt. The important new task for policy-makers is to identify, realistically, problems and issues where economic processes must be pressed to change and areas where it is possible to identify 'win–win' policies.

The need to introduce more economic incentives and instruments is often stressed. However, the mechanisms through which these instruments work is often not understood, or misunderstood. The tradition of applying economic instruments is rooted in the centrally planned system. In that system prices, particularly industrial input and intermediate product prices, were set centrally and if the production cost of a product was higher than the price the enterprises received subsidies. In such an economic system additional production cost items such as environmental fines had little impact on enterprise behaviour and no impact on the prices either. It was only a fairly complicated way of channelling state budget money for environmental purposes since revenues from the environmental fines were always earmarked for environmental spending. Furthermore, this mechanism gives the image that levying fines or charges on enterprises has no economic impact on households. Environmental policy-makers must consider in their 'fund raising' proposals two new issues inherent in the rules of a market economy. One new feature is that most of the enterprises face hard budget constraint therefore they would lobby against any new charges or fines. The second issue is that increased enterprise costs would change market supply and likely increase prices to be paid by the households.

Enterprise lobbying activity was very strong in the preparation of the new product charge legislation. Some features of collection and particularly the disbursement design of the product charges suggest that policy-makers resorted to extensive reimbursement in order to achieve consensus. The future use of such practice may be limited since it ultimately shifts unpredictable and possibly substantial financial burden to the households. Political concerns and debates continue to be a major part in putting forward any new proposals for environmental charges or other economic incentives but there should be an increased role for more sophisticated and enhanced economic and environmental impact analysis.

The transition to a new social and economic system brought about fundamental changes in the division of tasks and responsibilities across different levels and organizations of public administration and fundamentally new relationships across governments (both local and central), private enterprises and households. Uncertainties in the process and overlapping assignments have particularly adverse impact on two areas: (i) enforcement of environmental legislation; and (ii) evolution of a new financing system in which tasks and responsibilities are matched with appropriate financing mechanisms. Environmental protection financing is still overwhelmingly based on budgetary sources. Direct mobilization of private capital and non-budgetary sources is rather slow and at a low volume. Therefore it is important to assign utmost importance for clearer assignment of environmental responsibilities between different government levels, the private sector and households. Based on that, financial responsibilities can be assigned and an evolving but predictable system for self-financing environmental services can be designed.

Environmental protection faces changing institutional and political constraints but economic constraints are most binding. In order to achieve environmental quality improvements induced by policy measures it is necessary to move away from the practice of fairly stringent standards with weak enforcement. The objective of joining the EU can provide strong support for environmental protection. Nevertheless, it should be recognized that the 'mechanical' harmonization may have an opposite effect. If EC required regulation will be simply enacted with no regard to its economic feasibility it will reinforce the strict standard setting command and control approach with scattered and rather *ad hoc* enforcement. Such a system is certainly inefficient in an economic sense but its effectiveness in terms of environmental quality improvement is low as well. While negotiating for EU membership, Hungary is not in the position to discuss EU environmental standards and requirements. Therefore those must constitute Hungary's long-term environmental objectives properly incorporated into the legislation. However, attention and efforts should be concentrated on developing an economically and socially feasible 'compliance schedule' with detailed elaboration of necessary measures and actions for achieving the long-term objectives.

REFERENCES

A magyar gazdaság átalakulásának és fejlődésének programja (Restructuring and Development Program for the Hungarian Economy) (1991) Budapest.

Abel, I., Csermely, A., Kaderják, P., Pavics, L. and Fertő, I. (1993) *Environmental Implications of Economic Restructuring: The Case of Hungary.* Project Report, Budapest University of Economic Sciences.

Botos, B. (1993) Az átmenet iparpolitikája (The industrial policy of the transition). *Közgazdasági Szemle*, XL(6), 510–522.

Bowman, M. and Hunter, D. (1992) Environmental reforms in post-Communist Central Europe: from high hopes to hard reality. *Michigan Journal of International Law*, **13**(1921), summer.

Donáth, B. (1995) The use of foreign financial assistance in Hungary. In: G. Klaassen and M. Smith (eds), *Financing Environmental Quality in Central and Eastern Europe: An Assessment of International Support*. IIASA, Laxenburg, Austria.

Dzuray, E.J. (ed.) (1995) *The Emerging Environmental Market, A Survey of the Czech Republic, Hungary, Poland, and the Slovak Republic*. Regional Environmental Centre, Budapest, June 1995.

Environmental Action Programme for Central and Eastern Europe (1993) Document submitted to the Ministerial Conference, Lucerne.

Hungary's National Renewal Program (1990–1992). Budapest, 1990.

IEA (International Energy Agency) (1995) *Energy Policy in Hungary*. IEA.

Kaderják, P. and Lehoczki, Zs. (1991) *Economic Transition and Environmental Protection: Foreign Investment and the Environment in Hungary*. BUE Working Paper, 1991/5.

Koloszár, M. and Lehoczki, Zs. (1993) *Economics and the Environment: Case Study on Hungary*. Ministry of Finance, Draft.

Kornai, J. (1981) *Economics of Shortage*. North Holland.

KSH (1995) *Magyar Statisztikai Évkönyv 1994 (Hungarian Statistical Yearbook 1994)*. Központi Statisztikai Hivatal (KSH).

KTM (Ministry of Environment and Regional Policy) (1995) *National Environmental and Nature Conservation Policy Concept*. KTM, Budapest.

Kulcsár, L. (1991) Környezetgazdálkodás és lakossági tudat (Environmental management and public awareness). In: *Környezet és Társadalom*. KTM.

Lehoczki, Zs. (1993) *Environmental Administration in Hungary*. January 1993.

Lehoczki, Zs. and Morris, G. (1995) *Integrating Fiscal and Environmental Policy*. Ministry of Finance, Budapest.

Lehoczki, Zs. and Szirmai, V. (1988) *Környezetállapot és érdekviszonyok Ajkán (State of Environment and Interests Structure in Ajka)*. Final report. KTM.

Magyar Köztársaság Kormánya (Government of Hungary) (1992) *Tájékoztató az Országgyűlés részére az energia politikáról (Report to the Parliament on the Energy Policy)*. Budapest.

KTM (Ministry of Environment and Regional Policy) (1994) *Environmental Quality*. Budapest.

KTM (Ministry of Environment and Regional Policy) (1995) *Hungary: Towards Strategy Planning for Sustainable Development*.

Papp, K. (1995) *Fuel Prices*. HIID, Budapest, mimeo.

Tajthy, T. (1995) *Egy karbonadó bevezetésének hatása az energia felhasználásra, valamint az okozott légköri szennyezőanyagok kibocsátására (Impact of Introducing Carbon Tax on Energy Use and Air Emission)*. KTM, mimeo.

West Merchant Bank (1993) *The Privatisation of the Water Industry in Hungary*. Discussion paper, November, 1993.

World Bank (1990) *Hungary, Environmental Issues*. World Bank Background Paper. World Bank, Washington DC.

World Bank (1993) *Financing Water and Sewerage Investment in Hungary*. Discussion Note. World Bank, Washington DC.

World Bank (1995) *Hungary, Structural Reforms for Sustainable Growth*. World Bank, Washington DC.

CHAPTER 5

Latvia

Valts Vilnitis

5.1 BACKGROUND

Changes in society

Latvia is in transition from a centrally planned economy to a market economy. The problems, which are inherited from the period of the last 50 years and from the transition process, have a significant impact on the environment as well. The Latvian Government has applied for European Union membership, and a general trend of shifting from the Newly Independent States (NIS) markets to the Western markets can be observed. However, there is no general economic development programme at present. Only some overall directions have been sketched in the latest Government Declaration. More or less detailed programmes have been prepared only for energy production, transport, and forestry sectors. The future of other sectors, especially industry and agriculture, have not been planned yet. It is virtually impossible to make any prognosis about directions, the rate of development of several sectors, and the economy in general.

Some general trends still can be mentioned. There is a remarkable development of the service sector together with a rapid decline in industrial and agricultural production. There are severe problems in municipal infrastructure and public investments and the development of financial and real estate markets. Most of the large and medium-sized industrial enterprises are shut down or on stand-by. Only a few of them have managed to attract the necessary Western investments and have a hope of survival. The main reason for this is the loss of the NIS markets and low competitiveness in the West. The most significant exports are timber and food industry products. There is considerable income from transit operations, especially oil. The total activities of the transport sector are reaching 1980 levels.

A general decline in economic activities has been observed since 1989. Only in 1994 did production show some growth. The privatization of state owned

The Environmental Challenge for Central European Economies in Transition.
Edited by J. Klarer and B. Moldan. © 1997 John Wiley & Sons Ltd.

industrial enterprises only started in 1995. Thus, significant economic growth cannot be expected soon. Small businesses, particularly in the service sector and in the processing of agricultural products, are developing rapidly. However, lack of funding for investments and operational loans mitigates this process.

Privatization in agriculture is virtually completed. Still, declining markets and lack of funding decrease its efficiency. Most farmers are forced to carry out natural farming and produce most goods for their own consumption or exchange with their neighbours.

Most large industrial enterprises are not very attractive for potential investors as they are too big, often burdened with significant debts and need total modernization. They frequently need environmental clean-up. In particular, there are few chances for industries which were based on imported raw materials and spare parts such as RAF – the largest producer of mini-buses in the former Soviet Union. Frequently, investors use only the land, sometimes also the buildings, and virtually establish a new enterprise.

A drop in general and environmental investments, in particular, occurred during the last several years. In 1994 state investments only reached 2.6% of the state budget, and only 5% of the total state investments were allocated to environment.

The Government receives public pressure mainly on issues which are related to social questions like social security, health care, and public safety. The level of general environmental awareness is comparatively low.

The main attention of the State is directed to social security. The decline in the economy and market relations created a very hard situation for several groups of society. These are pensioners, families with more than three children and industrial workers who are on compulsory vacation because of enterprises which have closed or have experienced a decline in production. A large part of the population is living below the official poverty level.

Thus, social security, employment, wealth generation and quality of life have about the same level of importance. Most long-term issues such as education, health, and, unfortunately, the environment are not that high on the political agenda. It does not mean that environment as a governmental topic is neglected, but priority is given to short-term actions with obvious results.

On the one hand, it is very hard to do any kind of long-term planning because the situation is changing very fast, but on the other hand, it is very hard to plan any kind of environmental activities without general knowledge about what the directions are for the overall development of the country in the next 10 to 20 years.

State of the environment

The state administration structure and management system during the previous decades was inefficient, badly organized and irresponsible. Agricultural and forestry land was state owned. Their use was much less intensive than in Western Europe. Vast agricultural areas were abandoned, and yearly forest growth exceeded logging. As a result, inefficient land use allowed preservation of the natural forests, meadows and swamps where rich animal and plant populations are to be found now. Many of these species are on the edge of extinction in the western and north-western regions of Europe. Latvia, at present, can be proud of its comparatively untouched nature, vast forest areas, coastal areas with little construction, and low pollution background level.

The majority of environmental problems are concentrated in the so-called 'hot spots' – the largest industrial centres, transport hubs or territories formerly used by the Russian Army. Some environmental problems manifested in the country as a whole include: eutrophication and degradation of water ecosystems; excess usage of several natural resources; transboundary pollution; and accumulation of household and industrial waste. Several grave local problems have been created by excessive and, in many cases, chaotic urbanization. Housing areas were developed faster than municipal infrastructures. Thus, water supply, wastewater treatment and transportation systems are in bad shape and lack necessary capacity.

Natural Conditions

Latvia is situated in an area of transition from northern coniferous forests to broad-leaved forests characteristic of the moderate zone, from the oceanic climate to the continental one. Mixed forests are the type of zonal vegetation. Latvia's territory is crossed by many plant and animal species. During the last 13 000 years of change in natural conditions, every climate and vegetation period has left its marks. These include rare species, unique biotopes, and relict nature monuments.

About 42% of the territory (64 000 km^2) is covered by forests, more than 10% by bogs and fens. The forest area has slowly increased since 1939, when it was only 28% of the whole territory. Unfortunately, areas of grasslands during the same time have significantly decreased due to scrub, reed and forest overgrowth. These processes were mainly caused by very inefficient agriculture during the Soviet period. Traditional single farms were abandoned in 1945 and large, but badly managed, collective farms were created instead. As a result, the total area of agricultural lands decreased to about 2 million hectares.

Natural Resources

The main natural resources in Latvia are forests, peat, and minerals (sand, clay, gravel, gypsum). Most of the land is now in private ownership. About half of all the forests are state owned. In many cases timber, peat or minerals are the only immediate sources of income for the private owner or municipality, so frequently short-term interests force them to over-exploit these resources. As Latvia is poor in energy resources, there are serious plans to expand the use of peat, firewood, or wood chips for local heating needs. These plans raise concerns about the possibility of over-harvesting bogs and forests due to local fuel extraction.

The use of natural resources is regulated by a rather developed legal system. However, enforcement is not always in place. Thus, there are many cases of abuse. This especially can be said about forestry where official statistics show a much better situation than reality.

Level of Urbanization

As a result of urbanization (concentration of industry and population), 70% of the Latvian population (totalling 2.5 million) lives in towns. This includes the 34% who reside in the Rīga agglomeration. Industrial cities have become large sources of air pollution, wastewater and solid waste. That in turn has been the cause for deterioration of population, health and environment not only in these cities, but also outside.

Some of the most important cities should be mentioned (data from the Environmental Consulting and Monitoring Centre and the Environmental Protection Department):

– Rīga has a population of approximately 900 000 inhabitants. The population density is 2790 km^{-2}. Parks and gardens cover 22% of the city area. There are several large enterprises in Rīga which represent the chemical, machine building, electronics, timber processing and food industries. Large-scale cargo operations take place in the port of Rīga which is run by several private companies.

– Daugvapils is the second biggest city with 122 000 inhabitants. It has relatively few industries.

– Liepāja has 104 600 inhabitants. Built-up portions cover 69% and gardens and parks 12% of the city area. The only steel mill in Latvia is here as well as some large light industry and food processing enterprises. During the last 50 years, Liepāja served as a Soviet Navy base, and the port was used solely for military purposes. Since 1992 there have been serious attempts to re-build the port for cargo operations, but this process is still under way. A large part of the city was occupied by Soviet military personnel. It is now nearly empty. Infrastructure and services are very weakly developed.

– Jelgava has a population of 71 300. RAF, the only car and mini-bus producer in Latvia, is here. Another remarkable polluter is the skin processing factory which produces significant quantities of hazardous waste. Jelgava is also an important railway hub. There are also several

smaller industrial enterprises, mostly machine building, textile and food processing enterprises.
- Jūrmala is a resort city near Rīga on the coast of the Gulf of Rīga. It has a population of 59 600 inhabitants. The population density is only 600 km^{-2}. More than 64% of the city is covered by parks and gardens. The only large industrial enterprise, the Sloka Paper and Pulp Mill, has heavy economic problems. The city economy traditionally was built on tourism which came from the former USSR. In recent years, the flow of tourists from the East has decreased considerably and, thus, the income of the city inhabitants is relatively low.
- Ventspils has 47 500 inhabitants. This relatively small harbour city handles most of the oil and chemical transit operations in the country. The average income per capita in Ventspils in 1993 was twice as much as Rīga and four times higher than in Jūrmala. On the other hand, there are large storage facilities which can accommodate 60 000 m^3 of ammonia and more than 3 million m^3 of oil. Acrilonitrile transport reaches 41 000 t a^{-1} and methanol 435 000 t a^{-1}.

Other cities and towns are relatively small, but in some cases even these are heavily industrialized. For example, the town of Olaine has several large chemical and pharmaceutical enterprises with a considerable number of environmental problems mostly related to air pollution and hazardous wastes.

It also has to be mentioned that city transport, in particular private vehicles, raise growing concerns in larger cities as infrastructure was not in place even at the beginning of the 1990s. Cities can hardly accommodate the rapidly increasing number of cars.

Quality of Soils

Latvian soils can be considered relatively clean due to the low content of heavy metals and toxic organic compounds. In the majority of soils, the content of heavy metals is on the level of the natural background (see Table 5.1). A significant part – 23% (1990) – of soils are acid (pH < 5.6).

Two hundred and thirty thousand hectares, or 14.7% of the total arable land, are threatened by wind erosion. Three hundred and eighty thousand hectares, or 24.3%, are subject to water erosion (Ministry of Agriculture, 1994). Agricultural production influences natural ecosystems significantly, but at the same time it is an important branch of the Latvian national economy producing up to 40% of the GDP. Thus, sustainable development of agriculture is both the basis for economic growth and a way to preserve a natural and healthy environment. However, the country has not yet formulated a clear agricultural policy, and it is hard to envisage any rapid improvements in the near future.

Table 5.1   Content of heavy metals in soils of different mechanical consistencies in Latvia and The Netherlands (mg kg^{-1}, average values)

Metal	Sand		Sandy loam		Loam	
	Latvia	The Netherlands	Latvia	The Netherlands	Latvia	The Netherlands
Chrome	0.7–2.3	60.0	4.0–7.7	76.0	3.7–7.7	106
Nickel	1.0–3.2	15.0	2.1–4.2	23.0	3.0–5.0	38
Copper	0.9–3.3	19.0	2.3–3.4	24.0	3.0–4.5	33
Zinc	5.3–9.3	67.0	11–17	91.0	11–18	137
Cadmium	0.3–0.6	0.5	0.2–0.4	0.5	0.22–0.42	0.6
Lead	2.9–9.1	56.0	5.3–7.0	65.0	5.4–5.8	80

Source: National Environmental Policy Plan for Latvia.

Water Quality

There are 777 rivers in Latvia which have lengths in excess of 10 km. The total length of waterways is about 12 000 km. Besides these, there are over 3000 lakes, 802 of which exceed 0.1 km^2.

According to hydro-biological and hydro-chemical data, 85% of all the surface water is either slightly polluted or polluted. It has to be stressed, however, that this figure is based on the application of standards from the former USSR. After introduction of European Union compatible standards, the overall picture will not look so grave.

Eutrophication is the biggest problem. It is created by biogenous substances and is growing rapidly. The main sources of biogens are untreated municipal wastewater and agricultural runoff. In several places, water pollution includes accumulated dangerous substances such as heavy metals, chloro-organic compounds and oil products. However, it should be stressed that after 1990, both the amount of wastewaters and agricultural runoff decreased significantly. Accordingly, the total amount of pollution that was discharged into watercourses, mainly due to the start of operation of the Rīga Municipal Waste Water Treatment Plant (WWTP) and other WWTPs, as well as a decrease of industrial and agricultural production in the country (Figures 5.1 and 5.2), improved water quality. For example, the decrease of industrial wastewater discharge from 1991 to 1994 was 27%. Use of pesticides in agriculture in the same years fell to 90%. Runoff also was significantly decreased. Improved wastewater treatment ensured reduction of nitrogen by 54%. Phosphorus emissions were reduced by 30% from 1991 to 1994.

The situation in the Gulf of Rīga and the Baltic Sea should be stressed in particular. Relative isolation and significant discharge from rivers make the anthropogenous factor in the Gulf of especial significance.

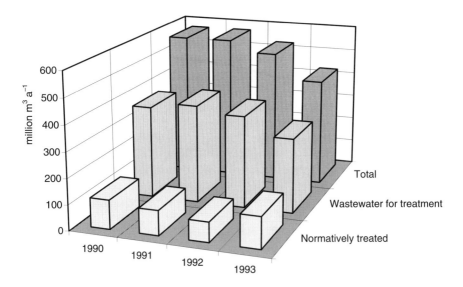

Figure 5.1   Amount of wastewater in Latvia, 1990–93. *Source: National Environmental Policy Plan for Latvia* (1995)

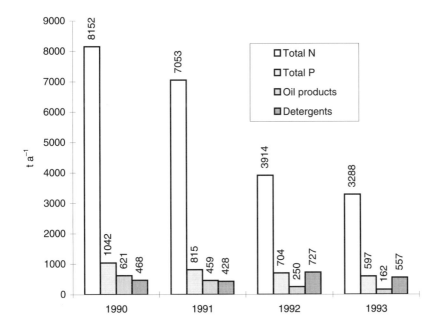

Figure 5.2   Amount of main pollutants in Latvian internal waters, 1990–93. *Source: National Environmental Policy Plan for Latvia* (1995)

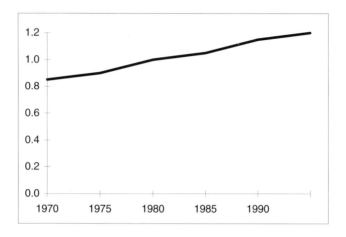

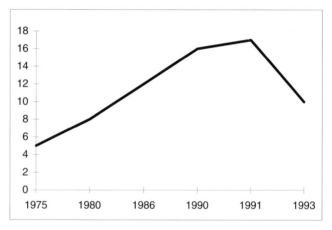

Figure 5.3   Total phosphate (top) and nitrate (bottom) concentration changes in the Gulf of Riga within a layer of 0–10 metres (μmol l^{-1}). *Source*: Hydrometeorological Agency, Sea Monitoring Centre (1994)

When analysing the figures over a period of several years, we can conclude that phosphorous and nitrogenous pollution in the Gulf of Rīga increased continuously since the end of the 1970s. Changes over the last years in concentration of biogenous substances in the Gulf of Rīga may be evaluated as a vivid example of water basin eutrophication (see Figure 5.3). As a result of eutrophication, the oxygen content of water decreases gradually during summer months. During the last year, a decrease of oxygen content in water during the summer was not observed. This change was caused by favourable hydrological processes, as well as a drastic fall in nitrate pollution. This can be

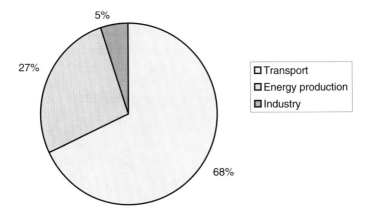

Figure 5.4 Distribution of air emissions in Latvia according to sources in 1993. *Source*: *National Environmental Policy Plan for Latvia* (1995)

explained by a decrease of nitric salts from rivers because of a fall in agricultural production.

Observations of phosphorus and chlorophyll concentration over the years show that eutrophication in the Gulf of Rīga continues. Analysis of changes in nitrogen and silicon concentration let us predict its cessation in the coming years if an influx of biogenous substances does not accelerate. Present results of hydrochemical and biological observations do not indicate further development of eutrophication in the Latvian Zone of the Central Baltics.

At present, the waters of the Latvian Economic Zone in the Baltic Sea are moderately polluted. Zones of Ecological Risk are coastal regions in the vicinity of river estuaries, locations of municipal and industrial wastewater discharges, locations of extraction of minerals and gravel disposal sites, as well as regions around ports.

Air Quality

The major part (65–68%) of air pollution is created by transport (see Figure 5.4). If compared with 1992, the 1993 air pollution caused by transport can be seen to have increased by 11%. This can be explained by the fact that the number of registered vehicles in Latvia for 1993 increased by 48 000. By the beginning of 1995, there was one private car per eight inhabitants in Latvia.

Pollution caused by district heating has also increased. This can be explained by changes in the fuel consumption pattern. More heavy fuel with a higher sulphur content is being burned and less natural gas. These changes are

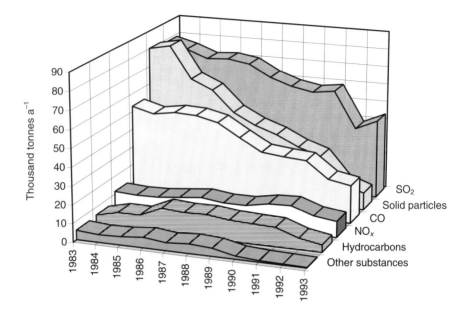

Figure 5.5   Emissions from stationary sources in Latvia in 1983–93. *Source: National Environmental Policy Plan for Latvia* (1995)

caused by a significant increase of gas prices and problems with gas imports from Russia.

Total exhaust emissions from non-mobile sources registered in Latvia in 1993 were created by sulphur dioxide (50.1%), nitrogen oxides (10%), carbon oxide (24%) and solid particles (10.8%). More than 200 other listed dangerous substances accounted for just 5%.

Figure 5.4 can be largely explained by the drop in industrial production in the beginning of the 1990s. The second reason for the prevailing role of transport in air pollution is the dominance of old cars in the total vehicle pool – the average age of a vehicle in Latvia is seven years. In 1994 the total amount of emissions increased again mostly due to shifts in fuel structure. During 1995, new car sales started to increase. The significant drop in emissions shown in Figure 5.5 reflects again a decrease in production. It is only partly the result of success of environmental policies.

Six of the largest industrial cities account for the largest part (57%) of non-mobile source pollution (Figure 5.6). In this aspect, Rīga stands out in particular – it alone accounts for close to one-third of total emissions in Latvia. It should be noted that these figures exclude small boiler houses and individual users. Calculations have determined that non-registered sources

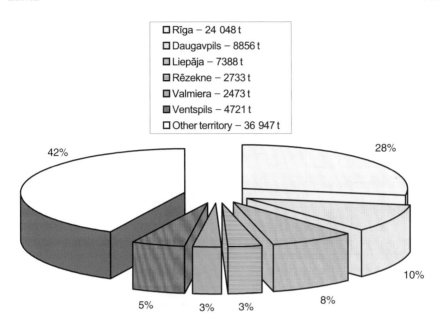

Figure 5.6   Air pollution from stationary sources in largest industrial cities in Latvia in 1993. *Source*: *National Environmental Policy Plan for Latvia* (1995)

increase emissions of dangerous substances in Rīga by another 18–20%, especially emissions of solid substances by 137% and carbon monoxide by 37%. An analogous situation can be expected all over Latvia.

Waste

According to data from the Environmental Protection Committee of the Republic of Latvia (1992), the Rīga agglomeration generates about 266 000 tonnes of household waste a year or 240 kg waste per person. In Latvia, the general figure is 200 kg. For comparison with other countries see Figure 5.7.

The present increase in household waste is mainly due to the expansion of Western imports and growth of the amount of packaging used. The second cause is the decrease of recycling and re-use of packaging created by changes in the state administrative and production structures during the last years. Latvia also lacks a developed household waste management system. It is considered that household waste is a responsibility of the municipalities. However, the central government is responsible for the development of relevant legal Acts which are currently lacking or underdeveloped.

Most landfills (there are about 500 in the country) are not built according to environmental requirements. At present, a national strategy is being developed for improving the overall municipal waste management system in the country.

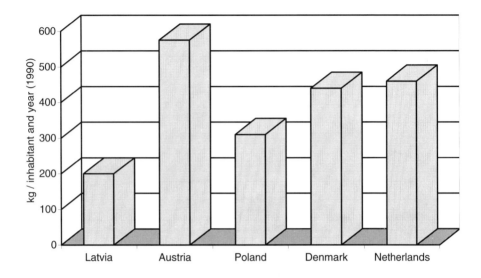

Figure 5.7   Amount of municipal waste generated in selected European countries (1990).
Source: *Europe's Environment* – *The Dobris Assessment* (European Environmental
Agency, 1995)

Hazardous wastes should be mentioned separately. Though they create just a small part of the total waste amount, these pose a particular risk to environment and human health. In 1989 and 1990 in Latvia, about 200 000 tonnes of hazardous waste were generated annually. In 1993, the amount generated was 42 000 tonnes. For comparison, Sweden generates about 500 000 tonnes and Denmark about 130 000 tonnes of hazardous waste per year. It is envisaged that a new national hazardous waste management system will start its operations in March 1997. A central storage facility is under construction. It is being based on a former Soviet missile base underground launching system. Also, a set of normative documents regulating the responsibilities of waste producers, the waste management company, central authorities and municipalities is expected to be completed by the same time.

5.2 INSTITUTIONS AND POLICIES – KEY PLAYERS

Existing environmental protection institutions and division of responsibility

At present the overall responsibility for environment on the national level lies with the Ministry of Environmental Protection and Regional Development (MEPRD). Other responsibilities of this Ministry include regional development planning, tourism policy, building policy and regulations, and municipalities (Figure 5.8). MEPRD was founded in August 1993. Before 1993, the

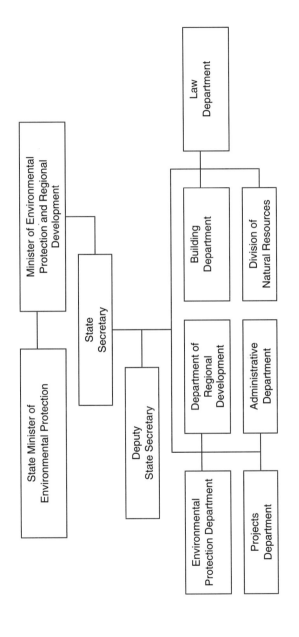

Figure 5.8 Organogram of the Ministry of Environmental Protection and Regional Development

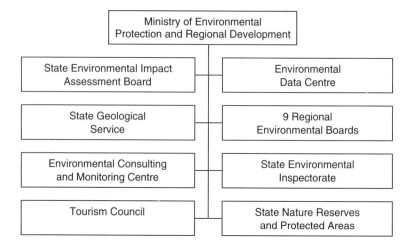

Figure 5.9   Environmental protection institutions in Latvia

highest environmental authority was the Environmental Protection Committee which was subordinated directly to the Parliament. MEPRD has seven departments. The Environmental Protection Department is solely responsible for environmental policy and regulations in the country. The Projects Department and the Law Department are largely occupied with environmental issues as well.

The Ministry has several subordinate organizations (Figure 5.9). Those directly related to the environment are: the State Environmental Inspectorate, the State Environmental Expertise Board (frequently referred to as the State Environmental Impact Assessment Board), the Environmental Consulting and Monitoring Centre, the Environmental Data Centre, administrations of the protected areas, and nine Regional Environmental Boards.

The Inspectorate and Expertise Board are responsible for enforcement of environmental regulations on the national level and guidance over the regions. The Inspectorate primarily deals with compliance with existing environmental laws and regulations for emission standards, land use and exploitation of natural resources. These are the responsibilities that inspectors have in the regions. The State Environmental Impact Assessment Board is responsible for the Environmental Impact Assessment (EIA) procedure. It ensures that the EIA Regional Boards cover the whole range of the Ministry's functions in the respective regions. The Regional Boards also implement EIA procedures for development projects of regional importance. Each of the Boards covers one to four administrative districts. A small Inspectorate is located in each district.

The Environmental Data Centre collects and processes data from the regions and other sectors. The Environmental Consulting and Monitoring

Centre is responsible for the design and function of national environmental monitoring, public dissemination of environmental information, analysis of environmental data, and advice on particular issues for all interested parties.

There are five state nature reserves which are handled by two administrative units – Slītere State Nature Reserve in the western part of Latvia, and Teiču State Nature Reserve in the east. There is also a North Vidzeme Nature Protection Complex, which is responsible for the management of multiple protected areas in the north of Latvia. At present, the only National Park in Latvia not subordinated to the MEPRD is Gauja. It is administrated by the State Forest Service (Ministry of Agriculture).

According to the existing legislation, responsibility for the environment and natural resources lies with the municipalities. State environmental authorities mainly have supervisory functions. However, at present most of the related activities are carried out by the authorities subordinate to the MEPRD. Several municipalities (mostly in the larger cities) have created their own environmental divisions. These deal with environmental issues on the local scale in co-operation with the Regional Environmental Boards and the MEPRD. A shift of responsibility to municipalities for most of these issues is not envisaged in the near future. The main reasons for this are the lack of sufficiently trained and skilled personnel, the lack of municipal funds and the high priority of development issues in the municipalities. There is a risk that if environmental management is totally given to municipalities, environmental issues in many cases may disappear from the agenda. However, negotiations will be going on with some municipalities about increasing their role in environmental protection from 1996 to 1997.

MEPRD, together with subordinate institutions, employs about 1500 people throughout the country. Most of them deal with environmental issues. However, the system still lacks the necessary capacity to carry out all its functions mostly due to the lack of adequately trained personnel, especially on the Regional Boards. Further expansion of this system is desirable but impossible at the time due to budgetary constraints. There are plans to establish an Environmental Protection Agency instead of several subordinated organizations, thus increasing the overall capacity and effectiveness of the system while at the same achieving some budgetary savings.

All the above-mentioned institutions are financed from the state budget. The natural resources tax provides some additional funding which is accumulated in the National Environmental Fund. The Fund is able to provide grants, subsidies and loans for environmental projects, both in the public and private sector. It cannot be re-allocated to anything but the environment.

Part of the natural resources tax is directed to the municipal environmental budgets, but there is no procedure for spending such moneys as yet. There is a hope that proper use of these resources might help to increase the interest of local authorities about environmental matters.

Influence of non-environmental institutions and sectors in decision-making

All draft legislative Acts related to the environment, before being sent to the Cabinet of Ministers for adoption, have to receive a positive response from the MEPRD. This is a usual procedure in Latvia, and the environmental authority is not the only one to be consulted. However, even drafts which are not approved by the Ministry may be adopted by the Government or Parliament (this happens very rarely).

The same procedure applies to all legislative Acts drafted by the MEPRD. Each ministry which has expressed its interest in the subject may send comments. These also must be considered by the MEPRD. The final decision is taken by the Cabinet of Ministers. In the case of laws, international conventions, and the state budget, the final decision is made by the Parliament. Usually the position of the MEPRD is fully taken into account except for financial issues where the budgetary situation has the determining role.

There are significant efforts to involve sectoral ministries in environmental decision-making, but due to various reasons (lack of interest, experience and manpower) they are not always successful. Municipalities are always involved in the particular activities which affect their territory. During the past few years, their involvement in local environmental decision-making gradually has increased.

Environmental policy

In spite of the fact that environmental institutions existed since 1988 in Latvia, a proper policy planning process was started only at the end of 1993 when the Ministry of Environmental Protection and Regional Development was established. The First National Environmental Policy Plan for Latvia (NEPPL) was elaborated during 1994 and accepted by the Cabinet of Ministers in April 1995.

The Plan was developed with assistance from the Dutch Ministry of Housing, Physical Planning and Environment (VROM). Advice was also provided by the Swedish Environmental Protection Agency (SEPA). VROM provided the Latvian team with advice on the policy planning process, facilitation for the planning workshops, and with resources needed for the organization of the planning. SEPA helped with advice on the integration of nature protection considerations in the document.

The main problem with starting an environmental policy development process in Latvia is lack of tradition and experience as well as weak or non-existent policies in other sectors. NEPPL was the first example of a participatory process aimed at the development of policy for an entire sector. People from different institutions such as the Ministry of Economy, Ministry of Finance, Ministry of Agriculture, and Ministry of Communication, were involved. Experts from academic circles and the Latvian Fund for Nature were invited as well. Due to the time constraints and limited human and financial resources, it was not possible to involve more than about 70 experts.

NEPPL is a document which contains broad environmental policy goals and specifies basic principles on which the environmental policy in Latvia is based. It lists priority environmental problems and provides a set of measures to tackle each of them. It gives an overview of the existing environmental policy instruments and indicates needs for the development of new ones. Its time frame is the next 20 to 30 years. NEPPL is considered to be a living document, and regular updating is planned.

In 1995 development of the National Environmental Action Programme (NEAP) began. It is fully based on NEPPL. The Programme is an action-orientated document with a maximum time frame of three years. The document was completed in October 1996. This is just the start of the so-called NEAP process in Latvia, one which will involve more and more different actors and target groups in the policy and operational planning process. NEPPL itself will be reviewed regularly – on the basis of The State of Environment Report, on one hand, and on the comments from the broad public and international community on the other. There are plans to establish an Integrated Environmental Management System in Latvia within the next few years. This would combine policy planning, operational planning, implementation monitoring and feedback from target groups.

Environmental law

Most of the legal Acts on the environment were developed and adopted after 1990 since the Environmental Protection Committee was established. However, practically all standards and norms, as well as methods for the calculation of the pollution discharges, are left from the Soviet period. Only in some cases were these slightly modernized and adapted to modern conditions. Since Latvia has signed an Association Agreement with the EU, there is a trend of shifting towards the European system of environmental standards. It is planned that the new system of standards and norms will be completed by 1998. However, enforcement of the system might take decades. The reasons for this are: the need for adjustments of the control capacity of environmental authorities including upgrades of laboratories, methodology and re-training of personnel. There also must be time allowed for polluters to be able to comply with the new requirements.

A short description of the most important environmental laws affecting the environment is given below:

- On *Environmental Protection* (1991) This is the umbrella law for environment. It includes: basic provisions and programmatic guidelines on the rights of the public to a quality human environment, information about the environmental situation, measures for the solution of many environmental problems, control in environmental protection, liability for environmental violations, international co-operation in environmental protection, allocation of competencies in environmental protection and other issues. This law is the first to refer to several new instruments for environmental protection: mandatory ecological insurance and ecological certification. However, mechanisms still have not been elaborated for implementation of these instruments.

- On *State Ecological Expertise* (1990) (also frequently translated as 'On Environmental Impact Assessment') This law governs the procedure for performance of state ecological expertise, goals, objects and other basic regulations. With this law, a very significant and powerful environmental policy implementation instrument was founded. At present, implementation of many plans, programmes, construction and reconstruction is not possible without a positive statement from state ecological experts.

- On the *Environmental Protection Committee of the Republic of Latvia* (1990) This was revoked in 1993 due to the establishment of the Ministry for Environmental Protection and Regional Development. It established a legal basis for many rights and instruments of environmental protection for state administration institutions (receiving information, the rights to visit any object in the territory of Latvia with the purpose of control of compliance with the environmental requirements, and suspending operations of industrial enterprises). Many ideas included in the law influence the activities of environmental protection institutions before adoption of new legal Acts.

- On *Natural Resources Tax* (1996) The law imposes a special tax for using natural resources in economic activity. Taxes are paid for extraction of minerals, discharge of pollution, water use, and sales of environmentally hazardous goods such as packaging and chemicals. Users must pay for discharge of pollution. Natural resources taxes are the most efficient economic instruments for environmental protection at present. The law first became effective in 1990. It was completely revised in 1995. The newest revision became effective in January 1996. The accumulation of environmental taxes in special earmarked funds is also envisaged.

- On *Hazardous Waste* (1993) This sets a procedure for operations with hazardous waste. The provision of the law on the ban of the import of hazardous waste has turned out to be very important for Latvia. It made it possible to refuse several large-scale projects prepared by Western companies that were envisaging treatment and storage of thousands of tonnes of hazardous waste in Latvia.

- On *Particularly Protected Nature Areas* (1993) This sets categories of protected areas and a procedure for their establishment and protection. The law also governs the rights of land ownership and land use in protected territories. It is the basis for retention of those areas during land reform and for the circumstances of re-establishment of private property.

- On *Radiation and Nuclear Safety* (1994) This is the first Latvian law in this field. It provides a procedure for operations with sources of ionizing radiation (licensing of entrepreneurial activity, procedure for receiving permits, as well as performance of other operations with sources of ionizing radiation) and other activities that ensure the preservation of human lives, health, property and the environment. Radioactive materials are used in medicine and laboratories. Radiation and nuclear safety is not a big issue for Latvia as there are no large-scale activities with radioactive materials in the country. The only research reactor in Salaspils is being prepared for shut-down within the next few years, and there are no other nuclear installations. The Ignalina Nuclear Power Plant in Lithuania is just 30 km from the Latvian border.
- *Regulation on Environmental Protection State Inspection* (1990) The regulation provides for the status and competency of Environmental Protection State Inspectors. The regulation also governs, in part, the procedure for receiving permits for use of many natural resources.

Relations with other sectors

At the moment relations with other sectors are not well developed, and those existing are mainly based on private contacts. This especially relates to agriculture, the energy sector, defence, and frequently, also, forestry. It is envisaged that the NEPPL will be able to provide a basis for integration of environmental considerations into sectorial policies. However, this is difficult to safeguard under conditions of the transition period. One of the most significant obstacles for developing a good dialogue between environmentalists and development sectors is low environmental awareness. Another one is a lack of clear representation of target groups: farmers, different industry branches, and consumers frequently do not have organizations which could be a dialogue partner for the environmental authorities.

Nevertheless, there are some exceptions. For example, since 1994 the MEPRD is rather successfully developing its relations with several munici- palities of different size and jointly preparing environmental investment projects. During this process, communication with the Ministries of Economy and Finance have been gradually improving, however, only in relation to particular projects. Overall, development planning still lacks sufficient co- ordination.

Education and information

There is no comprehensive system of environmental education developed in Latvia. There have been decentralized activities to introduce environment, nature protection and ecology in public schools, but these were random and did not influence the overall situation in the country.

Several academic institutions provide training in environment related subjects. The University of Latvia offers bachelor degrees in Environmental Science and Environmental Chemistry (two different courses – in the Faculty of Geography and Faculty of Chemistry – with respective emphasis). There is also a master's course in Environmental Science and Management, provided by the Centre for Environmental Science and Management at the University of Latvia in co-operation with the EU's TEMPUS programme. The Technical University of Latvia offers a bachelor's degree in Environmental Science as well, focusing mainly on pollution abatement measures.

Several bilateral and multilateral co-operation programmes give opportunities to young people who already work for environmental protection institutions to receive postgraduate training in environmental management and related fields.

Public information is not well developed at the time being. MEPRD has a Public Relations and Education Division, but it cannot provide for an effective system for public information alone. However, there are several specialized programmes on television. Information on the most important issues is regularly submitted to the mass media. The State of Environment Report was ready in May 1996 and has been made available to the public. A popular version of the Environmental Policy Plan, as well as a summary of the Environmental Action Programme, also will be published.

International relations and co-operation

Latvia is benefiting from a wide range of international co-operation programmes such as EU PHARE. It is participating in the Environment for Europe process and is a party to most global, European and regional environmental conventions. Particularly important is participation in activities under the framework of the Helsinki Convention. These are co-ordinated by HELCOM.

Co-operation among the three Baltic countries in the field of environment has improved since 1994. The most important events were signing the Trilateral Agreement between Estonia, Latvia and Lithuania in July 1995 and the foundation of the Baltic Environmental Forum (May 1995). The Agreement sets an overall framework for co-operation in the field of the environment among the three countries, sets basic principles for data and information exchange, solution of transboundary problems, and early warning in case of emergencies. The Baltic Environmental Forum is established as a framework for the exchange and dissemination of information and experience between Estonia, Latvia and Lithuania. It helps in the development of common policies by providing funding for various workshops and meetings. The Secretariat of the Forum is located in Rīga, and it has focal points in Tallinn and Vilnius. This project is supported by the EC DGXI and the German Umweltbundesamt (UBA).

With the creation of the Baltic Council of Ministers in 1994, the Committee of Senior Governmental Officials on Environment was established as well. The Committee is playing an increasing role in the co-ordination of environmental activities in the three Baltic governments.

Latvia has a number of bilateral co-operation agreements, e.g. with Germany, Estonia, Belarus, Sweden, and Poland. More are under preparation. Several Western countries provide Latvia technical assistance on a bilateral basis, in some cases without signing a formal agreement, but in the form of twinning arrangements (e.g. with the Ministry of Housing, Spatial Planning and the Environment of The Netherlands). There are also numerous environmental projects where Latvia receives considerable support from bilateral donors and international financing institutions in the form of grants and soft loans. These projects include technical assistance to national or regional environmental authorities, support to municipalities in improving environmental management, know-how transfer, capacity building and larger investment projects.

In the case of investment projects, a general rule is that Latvia contributes at least 20% of the funding to a project (without taking into account loans, which in the long run also are a form of domestic funding). These are usually shared between the national and local budgets. The rest constitutes a blend of loans and grant financing. At present, all large investment projects are in the area of municipal services: either wastewater treatment and water supply, or waste management.

There are several problems related to Western assistance. First, there is the problem of foreign technical assistance. For a number of years, priority was given to studies and content-orientated consulting. Latvia has well-educated local experts who can cover a wide range of know-how except in some very narrow and specific areas. At the same time, there is a considerable lack of process-orientated consulting because management skills are rather low in the country.

The second big issue has to do with investment projects. Governmental guarantees for loans are requested by most international financing agencies. This frequently creates a problem because of the strict governmental policy which strictly limits foreign loans.

Another big issue is co-ordination between the many different bilateral, multilateral and domestic processes. For the time being, this seems to be solved.

5.3 PUBLIC PERCEPTION AND PARTICIPATION

Public opinion about environmental issues

There are no reliable data available on the environmental awareness of the public. However, one can say that the general public understands that environmental issues are important, but they do not see any implications for them in everyday life. The last large campaign, which involved a broader range of people, took place in 1991. There was also a big campaign against the Daugvapils Hydroelectric Station and another against a proposed Metro building in Rīga. Since then the attention of the mass media and the public has shifted to other hot issues mainly related to the economy and well-being.

Sometimes a high level of public environmental awareness appears locally in relation to particular past uses (former Russian military sites) or planned activities such as the development of new landfill sites. In these cases, activists rarely search for any additional information on the issue. Unfortunately, the lack of public opinion research gives no opportunity for further analysis.

There are few NGOs in the country, and they are not organized well enough. Most of them are active on the local level and consist of 10–20 people. The only national grassroots environmental organization is the Environmental Protection Club which had exceptionally strong positions at the end of the 1980s. Since then they have lost their authority and organization. There is also the Latvian Fund for Nature which was developed mostly by nature protection scientists and does not have memberships. Its main function is to accumulate funds for nature management and research projects. There are also several international NGOs represented in Latvia such as the Worldwide Fund for Nature (WWF International) and the Coalition Clean Baltic (CCB).

State of public participation

Most legal Acts provide the possibility of public participation in the environmental decision-making process. For example, the Law on State Ecological Expertise envisages that all projects subject to an Environmental Impact Assessment (EIA) are made public by presenting them in open public hearings. However, a practical mechanism for such hearings is not sufficiently developed, and the activity of the public usually has been far beyond that expected.

The general public is able to make proposals for the legal documents prepared by the MEPRD. Unfortunately, up until now, activity has not been sufficiently high. For example, in 1991, when the framework Law on Environmental Protection was drafted, the draft was published in the newspapers and readers were encouraged to send their comments and proposals. However, only the Green Party and the Latvian Fund for Nature came up with proposals.

Some draft legal Acts regarding nature protection issues are being prepared by the Ministry of Environmental Protection and Regional Development. They are sent for harmonization not only to other ministries but also to several academic institutions and the Latvian Fund for Nature. Professional organizations usually provide rather good input into the preparation of the legal Acts, and their input has proven to be valuable in most cases. Communication between the MEPRD and the general public needs major improvement. Due to the lack of personnel, experience and funding, it is not very effective at the time being.

5.4 MAJOR PROBLEMS AND OBSTACLES

Major problems mainly have been listed above. However, it is worth while to list them briefly again. Most of them are the result of the complicated transition from a centrally planned economy to a market economy.

Ineffective administration could be mentioned as the most important problem. The situation has been improved considerably since 1990 by the creation of a new system of environmental authorities. However, most environmental institutions still need major improvements, relations between them should be clarified, and better interaction with sectoral ministries, and regional and local municipalities should be achieved.

Closely linked to the administrative shortcomings are *weakly developed information flows* between the environmental authorities of different levels and other actors. Monitoring systems exist, but different organizations that execute various monitoring programmes rarely co-ordinate their activities. In many cases, data or analytical conclusions are not reliable. Data storage and processing in the Environmental Data Centre (EDC) are rather well organized, but the EDC is responsible neither for the quality of incoming data nor for the analysis and interpretation of the data.

In spring 1995, an Environmental Consulting and Monitoring Centre was founded. Its objective was to put in order the National Environmental Monitoring System and to provide analysis and interpretation of environmental data.

There are plans to develop the Environmental Protection Agency (EPA) soon in order to solve part of the institutional problems and, simultaneously, to improve information flows. It is envisaged that several national authorities dealing with EIAs, data and information, and project management and implementation would be joined under the umbrella of the EPA.

Nationally, environmental issues lack proper *funding*. The environment, together with the energy and transport sectors, has been declared a priority area for national and international investments. However, current budget revenues cannot cover planned expenditure in these sectors. Donor financing is still available, but problems arise from the lack of domestic co-financing.

The budget of the MEPRD also is cut regularly due to national budgetary limitations. Thus, the role of the natural resources tax revenues and foreign assistance is gradually increasing. New natural resources tax legislation will lead to an overall increase of environmental expenditure in the country.

Lack of equipment and technical capability of the controlling authorities reflects the above-mentioned problems in the institutional and financial spheres. Several international technical assistance projects helped with some improvements, and environmental fines and fees provide certain funding for the Regional Environmental Boards. Still, the regional controlling authorities are not always in the position to react on every violation of environmental legislation. Moreover, significant increases will be needed to comply with the European Union environmental regulations.

Most *shortcomings in the legal system* are related to the need for a transition to the EU legal framework envisaged in the Association Agreement between Latvia and the EU. If most of the general environmental Acts should comply with the EU Directives, then the whole system of environmental standards and norms has to be completely changed (in some cases, only minor changes will be needed). The problem is not so much in accepting the EU standards, which as a rule are weaker than the current Latvian ones, but with enforcement. The latter is largely dependent on the institutional structure, analytical capabilities of the controlling authorities, and information flows which have been briefly mentioned above.

Problems with the *lack of public awareness, gaps in public information and the environmental education system* have been mentioned above. Most of the shortcomings are directly related to the overall transition process and can be solved only in line with the major changes in the society. However, MEPRD is involved in these issues by trying to broaden access to environmental information, ensuring a link between academic institutions and the environmental authorities, and supporting public awareness campaigns.

REFERENCES

Environmental Protection Committee of the Republic of Latvia (1992) *National Report to the United Nations Conference on Environment and Development*, Rio de Janeiro.
European Environment Agency (1995) *Europe's Environment – The Dobris Assessment.*
Ministry of the Environmental Protection and Regional Development (MEPRD) (1995) *National Environmental Policy Plan for Latvia* (in Latvian, English and Russian).

CHAPTER 6

Poland

Maciej Nowicki

6.1 PRESENT TRENDS IN ENVIRONMENTAL PROTECTION

Part of the inglorious legacy of the previous political system, with the centrally planned economy, is the disproportionately high level of environmental pollution in relation to the potential of Polish industry. This pollution is an inescapable side-effect of such features of the Communist economy as:

- The dominance of the most energy-consuming and anti-ecological branches of industry like coal mining and metallurgy.
- The establishment of huge industrial plants, usually located in heavily industrialized regions.
- Wasteful use of low-priced minerals, energy and water in goods production.
- Pricing of goods without any regard for their real production and social costs.

Moreover the State, as the exclusive owner of all industrial facilities, was not interested in introducing strict laws to enforce the construction of pollution-control installations. Annual expenditure for this purpose in Poland was only 0.2–0.3% of GNP, several times lower than in countries with a market economy. As a result, the natural environment had deteriorated progressively and, at the end of the 1980s, Poland was one of the most polluted countries in Europe.

After the change of the political system in 1989, there arose the basic conditions for reversal of this unfortunate trend. The system of a centrally planned economy was replaced by the free market economy, open for both internal and external competitions. Fast rising raw material prices and energy prices have caused an increase in effectiveness of their usage. The deep crisis of metal and military industries caused quick structural changes, diminishing the previous dominance of mining and heavy industry. The introduction of

The Environmental Challenge for Central European Economies in Transition.
Edited by J. Klarer and B. Moldan. © 1997 John Wiley & Sons Ltd.

Table 6.1   Emissions of the main air pollutants in Poland and OECD countries in 1989

	Particulates	SO_2	CO_2	NO_x	CO
		Emissions per capita (kg)			
Poland	63.1	102.8	38.9	3301	84.2
OECD	14.6	44.1	38.5	2651	142.6
		Emissions per unit area (tonnes/km^2)			
Poland	7.7	12.5	4.7	402	10.3
OECD	1.1	4.6	4.6	364	13.3
	Emissions per unit of primary energy (kg/tonne of coal equivalent)				
Poland	13.8	22.4	8.5	1029	18.5
OECD	2.8	8.2	6.4	637	24.0
		Emissions per unit of GNP (kg/'000 US$)			
Poland	15.0	24.5	9.3	787	20.2
OECD	1.8	5.4	4.0	290	14.3

Source: Ministry of Industry and Trade (1994).

fees and fines for emission of pollutants into the environment has forced enterprises to build up thousands of installations to serve the reduction of pollution.

All these factors have participated in the new positive trends in improvement of the state of the environment in Poland. These changes, which took place in particular sectors of environmental protection during the last five years, are described below.

Air protection

Late in the 1980s, air pollution in Poland was among the heaviest in Europe. In 1988–89 about 4 million tonnes of sulphur dioxide (SO_2), 1.5 million tonnes of nitrogen oxides (NO_x) and 2.5 million tonnes of particulate matters were emitted per year, making Poland the third biggest polluter in Europe (after the former Soviet Union and Germany). The emission of carbon monoxide (CO) and carbon dioxide (CO_2) was also very high (3.2 and 470 million tonnes per year respectively). These values were several times higher than in OECD countries, calculated per unit of GNP, per unit of energy consumption or per capita (Table 6.1). Concentration of SO_2, NO_x and particulates in the centres of all big cities substantially exceeded permissible standards. The most dramatic situation was in the Upper Silesia Region, in Cracow, and in the Legnica-Głogów copper mining area. In the so-called 'Black Triangle' in the Sudety Mountains, where the Polish, German and Czech borders meet, there are located 12 big power plants which burn brown coal (including one Polish power plant in Turów). None of these plants has

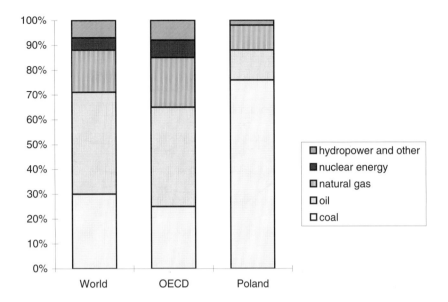

Figure 6.1   Primary energy use in Poland, OECD, and average in the world in 1990.
Source: Ministry of Industry and Trade (1994)

facilities for desulphurization of flue gases. As a result, the especially strong acid rains in this area caused the largest scale extinction of mountain forest in Europe. It was a striking example of the result of neglecting problems in air protection in Poland as well as in other COMECON countries from the previous political system.

The main reason for such high air pollution in Poland was the burning of coal for production of electricity and heat and obsolete industrial technologies without any devices serving environmental protection. Poland has a unique structure of primary energy consumption. As much as 76% of energy is produced from hard and brown coal (Figure 6.1). Almost 100% of electricity is produced from coal. Poland does not have any nuclear power plant and the resources of hydro-power are very limited. Hard coal is also the main energy source for provision of heat in all towns. There are still more than 9 million ceramic stoves in Poland and 1.5 million small boiler rooms burning coal and emitting about 1 million tonnes of SO_2 per year. Low chimneys constitute the main source of air pollution in the centres of all Polish cities.

The other reason for excessive air pollution was the waste of energy both in the industrial and residential sectors due to the low prices of all types of energy. According to the present assessments the level of possible energy savings in industry is equal to 30–40%. In the residential sector, it is 20–30% of the present energy usage.

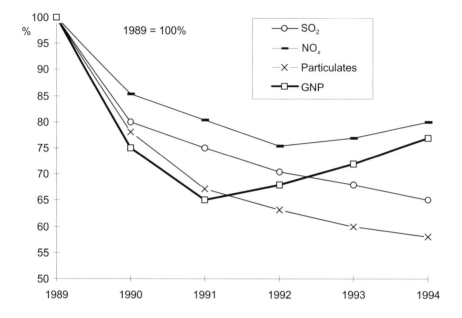

Figure 6.2   Changes in emission of SO_2, NO_x and particulates in Poland, 1989–94.
Source: Głowny Urzad Statystyczny (1991–95)

The third reason for air pollution, particularly important in highly industrialized regions like Upper Silesia, was the existence of old factories with outdated technologies. These are especially dangerous for the health of people living in neighbourhoods close to these factories. Such factories should have been closed down a long time ago, but the centrally planned economy had the principle of production for any price without calculation of real costs and impacts to health.

Presently the situation is changing dramatically. Figure 6.2 shows the changes in emission of the main air pollutants in Poland for the years 1989–94. It is especially interesting to compare these trends with the course of the Polish Global National Product in the same period of time. After a deep economic crisis in the years 1989–91, a substantial rise of the GNP and an increase in industrial production (4.5–5.0% of GNP per year) has been noticed for the last three years. Quite a different trend is observed in emission of the main pollutants. Emission of CO_2 has stabilized at a level 20–25% lower than in 1989. Emission of SO_2 and particulates has a decreasing tendency despite the rise of industrial activity. Especially spectacular is the reduction of particulate emissions (about 42% in comparison to 1988) and of SO_2 (in 1993 about 35% lower than in 1988).

There are many reasons for such good results. Most important are the following:

- Closing down of many obsolete factories.
- Saving of energy, mainly in the industrial sector.
- Reducing of coal and steel production by about 40%.
- Building up installations for coal purification, for reducing dust.
- The first desulphurization installations in power plants and boiler rooms.
- Converting from coal to gas as an energy source in many urban district heating systems.

It is worth mentioning that presently there are many desulphurization installations under construction in the biggest Polish power plants like Bełchatów, Turów, Rybnik, Jaworzno, Siersza, Opole and Konin. As a result of these initiatives, it can be expected that by the year 2000, reductions in emissions of SO_2 by 50%, in comparison with 1988, and reduction of particulates by 60–70% will be accomplished. It is realistic to predict that within 15–20 years, the level of emissions from these two pollutants will be comparable to the level in West European countries.

On the other hand, the problem of emission of air pollutants from the transport sector becomes more and more important. Poland has a very well developed railway system and only about 380 km of motorways. The railway still predominates as the primary transporter of goods and carries more than 60% of cargo. In big towns about 85% of people come every day to work using public transport. In the last five years, a trend to use many more private cars and trucks than public transport has been observed. During this time the number of cars has increased by 30% and the number of trucks by 20%. It is expected that this unfortunate trend will continue in the coming decade. It will be the main source of air pollution and noise in cities and along main roads and will cause substantial damages in environment and human health.

Water protection

In 1989 the state of water pollution in Poland was also very dramatic. As much as one-third of municipal and industrial sewage was dumped into rivers without any treatment and another 35% only after low efficiency mechanical treatment. Only the remaining 32% of wastewaters were treated to a satisfactory level. Almost 50% of cities did not have any wastewater treatment plants. Among the factories, only half of them had their own facilities for purification of sewage. The rest dumped sewage to the municipal sewage system. Their participation in the volume of sewage was an average of 30%, but in highly industrialized towns the share of industrial sewage reached a level of 80%. As a result of such negligence, the quality of Polish rivers was very poor. In 1989 only 5% of the total river length had potable water, and 35% of rivers had water so dirty that it could not be used even for industrial purposes.

A specific Polish problem is the high salinity of two main rivers, the Vistula and Odra, because waters which contain high loads of salt from the coal mines are dumped into them. The yearly amount of salt dumped into rivers is about 3 million tonnes, more than the entire salt consumption in Poland. This causes huge environmental damage and material losses in heating systems and in many factories using river waters. So far there is no good solution, not only in Poland but also worldwide. The main reason is that this process needs a big energy input and the price for the final product is very low. Several mines try to use different methods like reverse osmosis, recycling of saline waters or creation of emulsion with fly ash as a filling for the exhausted coal deposits. Probably there is no one universal method, and the set of technologies should be used depending on local conditions.

Another big problem is pollution of underground waters. The main sources of this process are as follows:

–   Communal sewage in rural areas
–   Substances used in the agricultural sector
–   Petroleum products from petrol stations and military camps.

An especially dramatic situation can be observed in areas where the Soviet Army troops had been stationed. In these huge areas, the soil is heavily contaminated and its purification will cost several billion dollars.

In the last five years, substantial progress in water protection in Poland has been observed. First of all, due to water savings, the amount of sewage produced by communes and industry decreased by 25%. At the same time, there was substantial progress in purification of wastewaters. The structure of wastewater treatment plants is shown in Figure 6.3. One can see that during 1988–93 the amount of unpurified sewage dumped into rivers decreased by 58%. Simultaneously, the volume of very modern and efficient biological and chemical wastewater treatment plants increased substantially. It is worth mentioning that the process of construction of such installations has also accelerated very much. The average time of construction is one to two years. Presently about 1000 wastewater treatment plants are under construction. Each year 350–400 of them are completed. If this investment boom is maintained, the problem of communal sewage will be solved by 2010.

Since almost the entire territory of Poland is within the Baltic Sea catchment area, the Sea receives most of the sewage which is dumped into rivers. In 1989 Poland contributed 40% of phosphorus contamination, 35% of nitrogen and 21% of organic matter to the Sea. About 50% of the population from the Baltic Sea basin lives in Poland, on about 40% of arable land in this region of Europe. Recently, the pollution of the Polish Zone of the Baltic Sea has

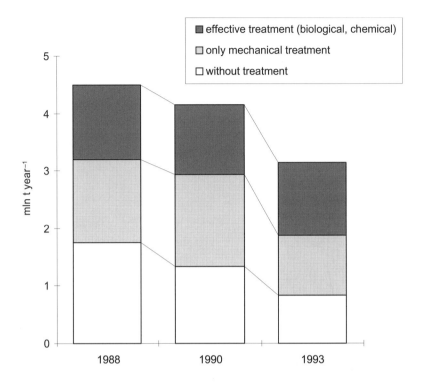

Figure 6.3   Structure of wastewater treatment plants in Poland, 1988–93. *Source*: Głowny Urzad Statystyczny (1991–95)

decreased substantially not only due to the smaller load of pollutants dumped into the rivers but also due to a decrease in the use of fertilizers and pesticides.

The majority of soil in the state owned farms is excluded from cultivation, and private farmers presently use manure instead of chemical substances on their small pieces of terrain (the average private farm has 6.5 hectares). It is necessary to be aware that this situation is only temporary and that Polish farming will be slowly changing towards the West European model. This can cause similar problems to the present ones in Holland, Denmark and Germany.

The only chance to avoid such a bad process of development is to promote the integrated model of farming in Poland with specialization in healthy food production. Especially Poland has the best conditions for such specialization – with its arable land split into a big number of small farm units, separated by lines of trees, bushes and balks. These farms are characterized by the rich fauna and flora which live symbiotically with agriculture as well as by the relatively clean and uncontaminated soil. At least two-thirds of Polish arable land has good basic conditions for healthy food production with a great benefit for the environment. It is still unclear if Poland will take advantage of this great chance.

During the next 15–20 years, the main problems of water protection in Poland will be the following:

- Shortage of good, potable water in many cities and regions of Poland.
- Desalination of effluents from the coal mines.
- Contamination of surface and underground water by the agricultural sector.

Waste management

In 1989 Poland did not have any efficient system for collecting and utilizing municipal waste. Every year 40–46 million m³ of such waste were dumped at disposal sites. There were more than 500 refuse dumps in towns and 1300 in rural areas. Simultaneously, there were more than 10 000 wild illegal deposit sites in woods or along countryside roads.

In 1989 industrial plants generated about 170 million tonnes of waste, and 43% of it was dumped into disposal sites. Included in this amount were about 3 million tonnes a year of hazardous waste because Poland did not have any installation for safe neutralization or utilization of this type of waste.

In contrast to the substantial progress in air and water protection during the last five years, there has not been any good change in waste management. During this period, the amount of municipal waste has increased by 20–30% because of the collapse of the system for collecting old paper and glass bottles. At the same time there has been increasing pollution from one-use packages. Until now no system has been established for recycling, composting and incineration of waste, and the number of illegal deposit sites is still growing. Although the first five composting plants already have been built (in Warsaw, Katowice, Zielona Góra, Kołobrzeg, Suwałki), their total capacity covers only 1% of communal waste produced in the country. In many towns the problem is rising dramatically because the old refuse dumps are full while there is a lack of acceptance for new sites to be placed in local communities. Thus, a breakdown in this sector can be expected in the near future because the problem of collecting municipal waste is growing really fast.

A similar situation also exists in the management of industrial waste. Despite a favourable tendency during the five years 1988–93 (Figure 6.4) of a fall in total volume, utilization stays at the same level. Almost 50% of waste produced in the industrial process is dumped on the deposit sites. Figure 6.5 shows the share of utilization of various types of industrial waste in Poland in 1993. It is worth noting that even waste from coal mines and fly ash from power plants is 50% deposited. There is an even worse situation in zinc and copper mining, where only one-third of waste is utilized. These numbers are

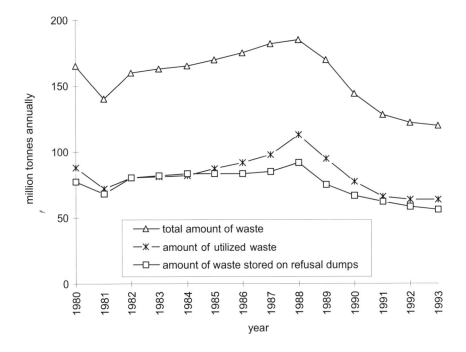

Figure 6.4   Industrial waste generated in Poland, 1980–93. *Source*: Głowny Urzad
Statystyczny (1991–95)

much lower than in the Western countries where 70–80% of industrial waste
finds many applications in the economy.

It should be stressed that so far about 2000 million tonnes of industrial
waste are collected at disposal sites throughout the country, but half of these
sites are located in the Katowice province, which is only 2% of the Polish
territory. Disposals consist of 800 million tonnes of waste from coal mining,
500 million tonnes of waste from the heavy metal industry and 300 million
tonnes of waste from the energy sector. At present there are the first successful
attempts of combining coal mine waste with power plant fly ash and placing
this product underground in the exhausted coal deposit areas. After the year
2000, this should be normal practice in the Polish coal mining sector.

Hazardous wastes (Table 6.2) are, and will continue to be for a long time, a
serious problem for Poland. Their yearly production reaches about 4.5 million
tonnes. Only 27% of this is utilized, and the rest is dumped at 802 deposit sites.
So far as many as 400 million tonnes of hazardous waste, mainly from the
chemical and pharmaceutical industry, have been deposited in such places. The
country does not have even one modern installation for safe neutralization of
such types of waste. This problem is urgent and needs more attention soon.

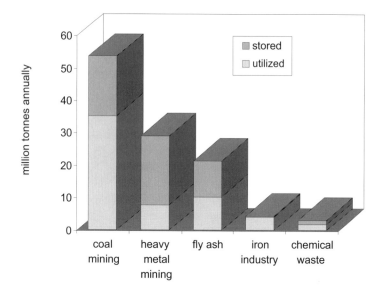

Figure 6.5   Utilization and deposition of various types of industrial waste in Poland in 1993. *Source*: Głowny Urzad Statystyczny (1991–95)

Table 6.2   Hazardous waste generated in Poland in 1993

Type of waste	Generated (× 1000 t)	Utilized (%)	Neutralized (%)	Stored (%)
Phosphorus-gypsum	1300	0.6	–	99.4
Waste moulding sands	950	48.0	–	52.0
Slurry and dust from metallurgical industry	500	86.2	–	13.8
Sludge from wastewater treatment plants	900	55.1	6.4	38.5
Waste from soda industry	600	22.8	–	77.2
Waste from painting production	120	5.0	–	95.0
Tars and acids from oil refining process	12	92.0	–	8.0

Source: Głowny Urzad Statystyczny.

It is necessary to state that the problem of waste management is still underassessed in Poland. It stays in the shadow of the two other main sectors of environmental protection – air and water protection. Although the situation is rather sad, many signs show that in the near future its importance will rise quickly, and that substantial progress will be noticed also in this area.

Nature conservation

Several features of Poland such as its location in the central part of Europe, its lack of natural barriers, both to the east and west, and its overlapping influences of continental and maritime climates have combined to create a set of habitats and species unique in the European scale. In the assessment of UNEP, the biological diversity of Poland ranks among the greatest in Europe, particularly regarding forest and bog communities characteristic of the Central European lowlands. There is a border of natural existence of 40% of European species of higher plants, 50% of vertebrates, 16% of birds, 22% of reptiles and 28% amphibians, running exactly through the Polish territory.

According to recent studies, the Polish flora includes more than 400 species of algae, about 5000 species of fungi and 5000 species of higher plants. Invertebrate fauna includes about 31 000 species, and vertebrate fauna more than 600 species. Especially rich is the world of birds (370 species) and mammals (98 species), including such rare species as wisent, wild bear, wolf, lynx and wildcat.

Despite the anthropogenic pressures of the twentieth century, biodiversity in Poland is still richer than in the West European countries. This is the result of an agricultural model with many features of ecological farming such as a mosaic field structure and a symbiotic relationship between wild nature and agriculture. Many national parks and biosphere reserves have been created due to the result of the work of many distinguished scientists and ecologists. Full protection of specially endangered species of plants and animals has also been introduced.

During the last six years, these ventures have been accelerated. Figure 6.6 shows the process of expansion of the protected areas in Poland. During this time, the area of national parks rose by 66%. It is expected that by 2010 as much as 30% of the country's territory will be covered by one of the diverse forms of nature protection.

Forest management plays an especially important role in nature protection in Poland. Forests cover 27.8% of the country's territory, but only 14% of them can be defined as natural forest. The rest constitute a pine monoculture which is especially sensitive to degradation due to anthropogenic pressure. Presently about 5% of forest areas are heavily endangered, especially in the southern and western part of Poland. The biggest damage is in the Sudety Mountains. There is a real disaster in an area of 13 000 hectares where spruce mountainous forests have become extinct because of acid rain. This is the so-called 'Black Triangle' area (see section on air protection above). Besides that there is big damage in forest ecosystems where the Soviet Army was stationed in different parts of Poland. The total damage is scattered over 60 000 hectares. Recultivation of these areas is one of the highest nature conservation priorities for the future.

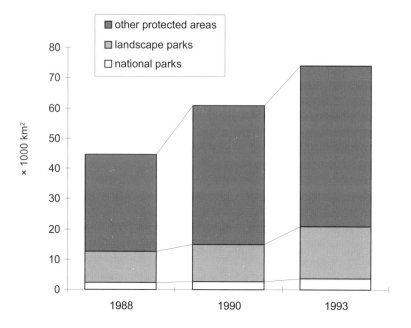

Figure 6.6   Protected areas in Poland, 1988–93. *Source*: Nowicki (1993)

A free market economy will create many potential dangers for the conservation of biodiversity in the most valuable areas. The threat is caused by changes in land use, consolidation of farmers' grounds combined with intensive agricultural practices, and big pressure to create a tourist infrastructure. There is a high probability that development in Poland will be similar to the process in the Mediterranean region where big hotels, roads and other facilities have destroyed the most beautiful and peaceful places.

A very promising attempt to systematically approach preservation of the most valuable areas with a simultaneous civilizing process is the idea behind Green Lungs of Poland (GLP). This area covers 22% of the territory of Poland (Figure 6.7) but is inhabited only by 8% of the country's total population. Half of the people from GLP live in villages. However, farming brings lower gains here than in other regions of Poland. This part of the country is very poor in mineral resources. Its biggest wealth is just nature. Forests cover 30% of GLP's territory; meadows, 32.4%; and wetlands, 8.5%. It is the location of 2000 lakes. These constitute 8% of the area. This beautiful macro-region is underdeveloped from the economic point of view. A market economy may create a great chance for improving the living standard of its inhabitants, but at the same time such an approach could lead to the deterioration of all its beauty in a long-term perspective. The only method to

Figure 6.7 'Green Lungs' of Poland

avoid such a threat is to develop Green Lungs of Poland in a sustainable way. Such thinking has been the basis for creation of a strong coalition composed of several ministries, nine voivods (regions) and local authorities to fulfil the idea, 'Environment and Development of Green Lungs of Poland'. There is a great chance that this slogan of the UN Conference in Rio de Janeiro will be realized for the first time just in Poland in a region whose territory is larger than Switzerland. It can be a first 'pilot project' of sustainable development on the local scale. The success of 'Green Lungs' could be multiplied in the future both in other regions of Poland and in Green Lungs of Europe created by Poland and the Baltic States.

Thus, generally speaking, Poland can offer to Europe the great value of its nature preserved in good shape and very rich in biological diversity. Present economic processes can diminish or even destroy all this beauty. Only with a good programme run over a long period of time can all the richness of nature be preserved. There is a need to establish new protected areas, pro-ecological forest management, and wider implementation of such innovative initiatives like the Green Lungs of Poland. This is the only way to avoid brutal anthropological pressure in the most valuable terrains of the country.

6.2 KEY INSTITUTIONS

The national level

Environmental protection is a multi-sectoral issue with responsibility divided between many central and regional authorities and all types of users of the environment (enterprises, society). Nevertheless, the special Ministry of Environmental Protection plays a crucial role as a main co-ordinator of efforts for the protection of the environment at the national level. In Poland it is the Ministry of Environmental Protection, Natural Resources and Forestry. As the name of this institution indicates, the Ministry is responsible not only for environmental protection but also for nature conservation, proper forest management and proper management of all mineral resources. This unique construction is very good in terms of creating a comprehensive strategy of sustainable development of Poland. Figure 6.8 presents the structure of this institution.

The main Ministerial duties are as follows:

- Elaboration and implementation of the National Environmental Policy.
- Formulation of laws, regulations and quality standards in the environmental protection area.
- Control of observance of ecological laws by users of the environment (through the Inspectorates for Environmental Protection).
- Co-ordination of the biggest environmental investments having crucial importance at the national level.
- International co-operation in the sphere of environmental protection.

In the framework of this Ministry, there also is included the National Commission for Environmental Impact Assessment, established in 1989, which consists of officials, researchers and representatives of non-governmental organizations. This Commission takes into consideration only the biggest new investment projects with potentially negative environmental impacts. For the first three years of its activity, it has assessed 17 projects and rejected 8 of them. Three similar commissions exist at the regional level.

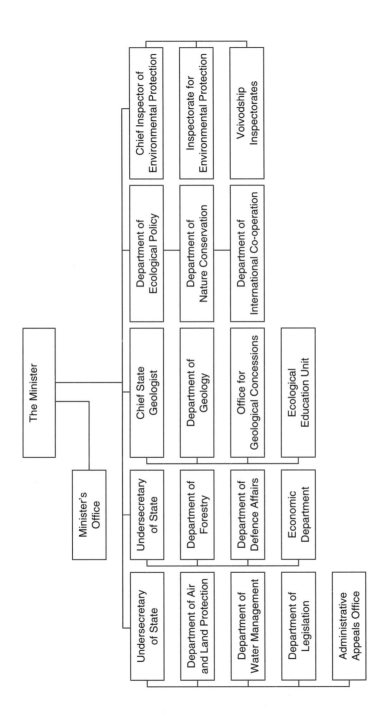

Figure 6.8 Structure of the Ministry of Environmental Protection, Mineral Resources and Forestry in 1995

Table 6.3   Responsibilities of the central institutions in the environmental protection area

Ministry	Responsibility
Ministry of Environmental Protection, Natural Resources and Forestry	The main co-ordinator of environmental policy (see text)
Ministry of Industry and Trade	National policies in industrial and energy sectors
Ministry of Housing and Physical Planning	Municipal water supply and sewage systems Management of municipal waste Land-use planning
Ministry of Finance	Financial instruments in ecological policy
Ministry of Privatization	Environmental audits in the process of privatization of enterprises
Ministry of Agriculture	Import and distribution of pesticides
Ministry of Health and Social Welfare	Monitoring of environmental standards connected with health protection Control of food products
Ministry of Labour and Social Policy	Occupation safety
Ministry of Transport and Maritime Economy	Transport policy, ecological standards in establishment of new roads Environmental issues related to the Baltic Sea
Ministry of Education	Ecological education in the entire school system
Scientific Research Committee	Financing researches in the environmental protection area
Committee of Sport and Tourism	Development of ecotourism
Central Planning Office	Strategic programmes of sustainable development of the country

Many other ministries and central offices play an important role in relation to environmental policy. Table 6.3 shows very clearly that almost the whole Cabinet is involved in creation and implementation of this policy. Thus, progress in environmental protection in Poland strongly depends on the quality of work of particular ministries.

Unfortunately, many of these institutions do not pay enough attention to this subject. They do not have special departments but only individuals, who are not active in fulfilment of duties, belonging to the scope of each given ministry. These individuals are link officers for connections with the Ministry of Environmental Protection and for elaboration of opinions concerning new regulations worked out by the specific institution. As a result of such an approach, only the Ministry of Environmental Protection is identified by the society as an exclusive office responsible for environmental issues.

This problem becomes increasingly important for the further development of Poland according to rules of sustainable development. The close and active co-operation of all economic, social and ecological sectors is the precondition for this process. The awareness of the need for such co-operation in the establishment of a comprehensive strategy of development of Poland during the next 20–30 years is not very high so far. Creation of the National Commission on Sustainable Development in November 1994 was the first step in this direction. The Minister of Environmental Protection is a chairman of this Commission, while other ministries are represented by Under-Secretaries of State. The Commission has the task of advising the Cabinet on all questions concerning strategic plans and giving opinions regarding the Polish reports to the UN Commission on Sustainable Development. Thus, the role of this Commission is rather limited. However, there are hopes that in the future its importance will be at least at the same level as that of the Economic Committee of the Council of Ministries which prepares all economic documents for the Cabinet.

The regional level (voivodships)

Poland is divided into 49 regions called *voivodships*. In each of them there is a Department of Environmental Protection which plays a crucial role in the implementation of National Environmental Policy at the regional level. These departments are responsible for decisions such as the following:

–   Permits for establishment of new factories and infrastructure facilities.
–   Bans on activities dangerous for the environment.
–   Permissible emission limits for each source of emission into the environment.
–   Infliction of penalties in the case of excessive emissions or discharges.
–   Collection of fees paid by enterprises for emissions and discharges.

In many issues, the regional Department of Environmental Protection makes the final decision on violations cited by the Regional Inspectorate of Environmental Protection. All problems connected with nature conservation at regional level belongs to the Conservator of Nature for the proper voivodship.

The local level (gminas)

Gminas are the smallest administrative units in Poland. There are more than 2100 of them in the country. Each gmina is ruled by a government chosen in a direct election by its citizens. Self-government of a gmina does not depend on the State or regional government, but it is obliged to realize the national environmental policy in the frames of its competency. Local self-government is responsible for:

- Providing a supply of good quality drinking water in a sufficient quantity.
- Establishment and maintenance of municipal sewage.
- Organization of a municipal waste management system.
- Elaboration of land-use plans and control of their observance.
- Establishment of green squares and parks and taking care of communal green areas.

These tasks are the result of a deep decentralization of the administrative system in Poland made between 1990 and 1991. This reform was one of the biggest achievements of the democratization of the political system. It turns out that active people from local communities fulfil their duty in their own city or village much better than clerks in central offices.

Agencies connected with environmental protection

An especially important role in implementation of the National Environmental Policy is played by the State Inspectorate of Environmental Protection (SIEP). This agency depends directly on the Minister of Environmental Protection, Natural Resources and Forestry. Its two main duties are:

- Controlling of observance of ecological laws and regulations
- Nationwide monitoring of the state of the environment.

After deep reshaping of this institution in 1991, its inspectors enjoy broader rights for control of the production process from the environmental point of view in each type of enterprise. In the case of violation of environmental standards, inspectors may issue penalties and close down a technological line or even the entire factory if it is responsible for causing a health danger. Between 1991 and 1993, seven enterprises were closed down completely. In the other 25 enterprises, the most dangerous parts of technology were cancelled. SIEP pays special attention to the 80 most dangerous factories in Poland – officially published in 1990 by the Minister of Environmental Protection. As a result of its strict control for the last three years, the following results have been achieved: the reduction of SO_2 emission by 40%, particulates by 60%, waste by 40% and load in sewage by 70%. This spectacular example shows how good effects can be reached as a result of regulation of the most polluting enterprises. Such actions have created quick and efficient improvement of environmental conditions in the highly industrialized areas.

SIEP is also responsible for organizing and operating the national monitoring system which measures the concentration of air, water and soil pollutants (background measurements) at the local (i.e. town, gmina), regional and national level. It also gives certificates for the measurement equipment allowed for use in Poland (i.e. automatic meters) and decides about unification of methods and standards of measurement used in the monitoring system.

For the last three years, SIEP has received a lot of very modern, automatic equipment both for control and monitoring purposes – a very optimistic sign. Now the level of SIEP's service provisions is really high, no lower than those of similar institutions working presently in Western Europe. Modern equipment is a basic condition for increasing the better functioning of this institution that is of crucial importance in the entire system of environmental protection in Poland.

Other central institutions and agencies that deal with environmental protection are the following:

- State Sanitary Inspectorate, dependent on the Ministry of Health and Social Welfare (responsible for measurements of environmental pollution from the health protection point of view).
- State Forest Administration (manages the state forests which constitute 78% of the country's forested land).
- National Parks Administration which co-ordinates management in 20 Polish National Parks.
- Research institutes dependent on the Ministry of Environmental Protection, Natural Resources and Forestry such as the State Institute of Meteorology and Water Management, State Institute of Geology, Institute of Environmental Protection, Institute of Ecology of Industrial Terrains, State Institute of Forestry.
- Seven Regional Water Management Boards which should have in their duties the water management in seven natural basins. The tasks of these Boards can be fulfilled only after the Polish Parliament adopts the new Water Law which has been in the legislative process for a long time.

6.3 TOOLS FOR ECOLOGICAL POLICY

National environmental policy

The necessity of working out the strategy of environmental protection in Poland arose for the first time during the 'round table' negotiations between the Communist Government and the still illegal 'Solidarność' in March 1989. This document was prepared in 1990 and adopted both by the Government (November 1990) and by the Polish Parliament (May 1991). It contains the basic conditions for environmental protection until 2010 in a new, democratic political system with a market economy. For the first time there were formulated several principles of sustainable development policy:

- Necessity of reconstruction of the entire legal system.
- Decentralization of the decision-making process and regionalization of environmental policy.
- Implementation of the 'polluter pays' principle.

- Control of pollution at the source.
- Public participation in the decision-making process.
- Broad ecological education of the Polish society.
- International co-operation.

The National Environmental Policy determines the short-term (3–5 years), medium term (10 years) and long term (20–30 years) priorities. In the first three to four years (until 1995), the main health threats had to be eliminated as well as the main negligences compensated.

The medium-term priorities are to create a substantial reduction in the emission of main pollutants to air, water and soil and to change environmental standards to meet those of the European Union by the year 2000.

The long-term priorities provide for implementation of sustainable development of the Polish economy together with the desirable state of the environment in 20 to 30 years.

The legislative, administrative and economic tools for achieving these ambitious goals are described in the National Environmental Policy. There are also directions concerning the enforcement of law, ecological education and international co-operation. Such a concrete basic document, showing a long-term and broad perspective, is quite unique both in Polish Government acts and in all post-Communist countries of Central and Eastern Europe.

In 1995 the Polish Parliament adopted the report as the realization of the first stage of implementation. Although not all short-term goals had been fulfilled, aims from several other fields foreseen for the year 2000 were already achieved. The general appraisal of the progress in implementation of environmental policy was positive. The same impression was concluded in the comprehensive survey titled *Environmental Performance Review* (OECD, 1995). It was concluded, in this valuable and objective report, that 'significant environmental improvements have been achieved, largely as a result of the contraction of economic activity and restructuring of the industrial and energy sector, but also as a result of environmental policies adopted and implemented'.

It is worth stressing that Poland has been selected as the first non-OECD country for conducting such a comprehensive review. Recommendations of this study are very valuable for further implementation of ecological policy in terms of Poland becoming an OECD member.

Reforms of the ecological law

Effective activity in the environmental protection area in a democratic country is based on good laws. Due to the complexity of problems connected with the environment, it is very difficult to elaborate the Ecological Code describing in one comprehensive Act all aspects of environmental protection. Such an Act would be the best solution, and it should be done in the future.

In the meantime, in the transitional phase of the country, there is a mix of old Acts issued in the period of the centrally planned economy and new ones adopted after 1989. Old Acts and regulations are to some extent anachronistic and should be replaced as quickly as possible. Among them there is the basic ecological Act entitled *Statute on the Protection and Shaping of Environment* issued in 1980.

Since 1989 the Polish Parliament has adopted several new Acts:

- Statute on the State Inspectorate for Environmental Protection (1991).
- Statute on Forest Management (1991).
- Nature Conservation Act (1991).
- Statute on Land Use Planning (1994).
- Statute on Mining and Geological Concessions (1994).
- Decree on Environmental Impact Assessment (1990).

Unfortunately, despite the six years of deep reform in all spheres of life in Poland, the process of modernization of ecological law is still far from being completed. This is a real obstacle for effective functioning of institutions responsible for environmental protection at all levels. It is a pity that very intensive and even pioneering efforts in this respect made in the years 1990–91 have been stopped lately to a great extent. Several important Acts have been waiting for adoption for years. Among them are the following:

- The Water Law.
- The Waste Management Act.
- Statute on the Extreme Threats for Environment.

The Air Protection Act and new version of the Statute on the Environmental Protection are still under preparation. Hopefully, in the near future, the outlines of mentioned regulations will be completed and adopted by Parliament. In such a way, a good solid foundation will be established for an effective, full-scale implementation of the environmental policy.

Financial instruments

It is obvious that the basic precondition of effective environmental protection lies in sufficient financial resources devoted for this purpose. One of the main reasons for huge neglect in this area from 1970 to 1980 was that only 0.2–0.3% of the GNP was invested during these years for environmental protection (Figure 6.9). This share rose from 0.6 to 0.8% of the GNP from 1985 to 1989, but it was still not enough to reverse the unfavourable trends which led to environmental damage in heavily industrialized regions of Poland. The experience of many countries has shown that an outlay of less

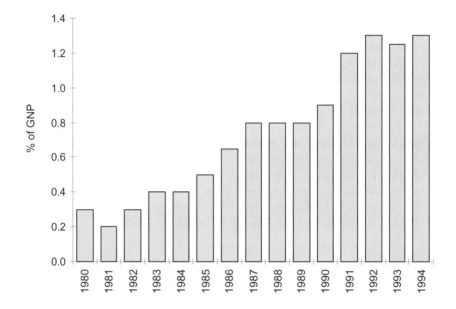

Figure 6.9   Annual expenditure on environmental protection in Poland, 1980–94.
Source: Nowicki (1993)

than 1% of GNP is not enough to stop a progressive deterioration of the environment. The required minimum is at the level of 1.3% of GNP.

After 1989, despite a deep economic crisis, environmental protection was the only sector which showed a funding increase. Expenditure in this sector very quickly reached 1.3% of the GNP (see Figure 6.9), an amount characteristic of rich Western countries. It also should be stressed that an additional 0.7% of the GNP has been devoted every year for water management.

The investment boom in the environmental protection sector is clearly visible in the background of investments in all economic sectors presently made in the country. Devices directly serving environmental protection consist of as much as 6.7% of all investments and create jobs for about 300 000 workers.

All these achievements were possible because of the application of quite innovative financial mechanisms from 1989 to 1991. These are presently working very well and are only slightly affected by political changes so typical for all 'countries in transition'.

The first substantial change, in comparison to the centrally planned economy, was the liquidation of state protection of state owned enterprises which caused environmental damage. The activity of the State Inspectorate of Environmental Protection has led to the closing down of many outdated factories and to the real enforcement of ecological law. Many other enterprises have realized that they are in fact obliged, according to the basic rule that a polluter must pay, to install devices which protect the environment. In 1990 the Ministry of Environmental Protection issued an official list of 80 enterprises which were the most dangerous nationally for the environment. These firms were obliged to reduce drastically the emission of main pollutants during three to four years. The voivods issued similar regional lists of 800 enterprises responsible for causing environmental damage.

The second important step in the process of improving the environment has been changes in the communal sector – one of the main sources of environmental pollution. Since 1990, local authorities have become responsible for water supply, sewage systems together with purification of sewage, provision of heat through district heating systems, and communal waste management. Decentralization reforms in 1990–91 gave a better basis for bigger spending from commune resources than the system of centrally steered subventions.

It is obvious that in the transitional period it is impossible to put into force the 'polluter pays' principle because of the huge neglect of the past. Thus, at present, additional mechanisms should be used to support the work of polluters in reducing environmental pollution which they cause. The most powerful mechanism is connected with introducing fees, paid by all users of the environment, for emission of pollutants to air, water and soil. These fees are collected by ecological funds. About 50% of them go to the National Fund for Environmental Protection and Water Management (NFEP), established in 1989, and the rest go to the regional ecological funds.

Since 1993 a small part of fees also have been received by gminas. Money from this source is spent for supporting the most important pro-ecological investment projects making it easier for them to be implemented. Thus, a kind of optimization of expenditure from the limited financial resources which are available in the country for environmental protection has been created. It is important to stress that the average payments made by enterprises amount to a fraction of a per cent to several per cent of production costs and constitute a part of the product's price. This way the whole Polish society pays a special tax for improvement of the environment just by purchasing goods. The enterprise, which exceeds the permissible level of emission, must pay a penalty ten times higher than a fee. Such a penalty is really galling because it must be paid from the profit and cannot be included in the price of goods.

For the last three years the participation of environmental funds in total expenditure in Poland has ranged from 40 to 58%. It is expected that until the year 2010 this source of financing will stabilize at 30–40% of the total. Table 6.4 presents the total expenditure on the environmental protection sector in Poland from 1991 to 1993. During this time, the share of enterprises' input in total expenditure was only at the level of 20–30% because many of them made big investments for modernization of technologies. Their participation in environmental protection investments should rise in the near future up to the level of 30–40%.

Table 6.4   Expenditure on environmental protection in Poland, 1991–93

	1991	1992	1993
Total expenditures (in million US$)	757	878	889
Sources of funding (%)			
– environmental funds	40	58	47
– enterprises	30	20	25
– communes	20	13	16
– state budget	5	5	6
– foreign aid	5	4	6

Source: Nowicki (1995b).

The third biggest source of funding for environmental protection is the commune (15–20% of the total spending comes from this source). Here it should be pointed out that the projects in the municipal sector are mainly supported by ecological funds in the form of soft loans. Local authorities are the main clients of the National Fund for Environmental Protection and for regional ecological funds. As these loans have to be paid off, the real level of the share of the communes in the expenditure on environmental protection in Poland can be estimated as 65–70% of the total amount. In the future their participation probably will be stabilized at 50–60% of the country's total expenditure in the environmental protection sector.

The state budget participates only 5–6% in financing environmental protection because of a specific Polish mechanism. This money is primarily spent on state agencies which are involved in enforcement. The budget provides for nature conservation projects, particularly those involved in maintenance of national parks. It should be noted that this source provides additional measures for water management, including construction and maintenance of reservoirs, river management and flood protection works.

The last 5–6% of expenditure on environmental protection comes from the official bilateral and multinational aid of foreign governments and international institutions (annually US$30–50 million). Table 6.5 presents the main foreign sources of funding in the form of donations on environmental protection in Poland from 1990 to 1994. Out of the total US$298 million, about US$240 million were spent in the last five years.

The majority of expenditure within the framework of the PHARE Programme and bilateral assistance was devoted to master plans, feasibility studies and business plans as well as to the training of Polish specialists and equipment of inspectorates (Figure 6.10). Only one-third of these funds were spent on construction of installations which directly reduced environmental pollution.

The structure of expenditure arising from the Polish debt-for-environment swap is completely different. About 90% of the money coming from this

Table 6.5   Main foreign sources of funding on environmental protection in Poland, 1990–94

Source of funding		Million US$	
PHARE Programme	(European Union)	92.1	(Five Phases)
Bilateral aid	(11 countries)	151.4	
including:	USA		36.3
	Denmark		35.5
	Germany		28.5
	Sweden		22.0
	Holland		9.0
Debt-for-environment swap			
(various forms)		44.5	
including:	USA		19.5
	France		1.5
	Switzerland		1.0
	Germany		15.0
	Finland		7.5
	TOTAL	288.0	

Source: Nowicki (1995b).

source is spent on construction of installations. The debt-for-environment swap is an official foreign debt and an original Polish initiative unique throughout the world. In 1991 Poland's official debt towards the Paris Club amounted to about US$32 billion. The Paris Club decided to cancel 50% of this debt under the condition that the rest would be repaid by Poland in yearly instalments until 2010. The Polish side made the proposal that an additional 10% of the debt (beyond 50%) could be converted into zlotys and spent on environmental protection projects. The Paris Club has created such an opportunity for all its members. This means that from this source Poland could receive a support up to US$3 billion until 2010. So far only three countries decided to use the debt-for-environment swap idea:

−   USA (10% reduction of the debt, US$370 million).
−   France (1% reduction of the debt, US$52 million).
−   Switzerland (10% reduction of the debt, US$50 million).

As a result, in 1992, the Minister of Finance established the ECOFUND, an institution responsible for management of these moneys. ECOFUND supports projects (in the form of donations) in four areas of international significance:

1.   Reduction of emission of greenhouse gases (CO_2, methane, CFCs).
2.   Reduction of gases causing acid rain (SO_2, NO_x).
3.   Purification of the Baltic Sea.
4.   Preservation of biological diversity.

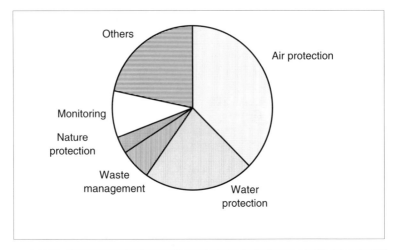

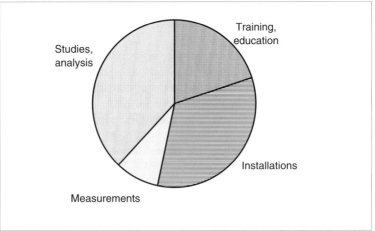

Figure 6.10   Aims of financial support of PHARE and bilateral aid on environmental protection in Poland, 1990–94. *Source*: Nowicki (1995b)

In the first two years of its operation ECOFUND supported 60 projects with a total amount of US$22 million. Since 1995 it has had at its disposal US$30 million yearly. An important task of the ECOFUND lies in promotion of the best technologies in the environmental protection area on the Polish market. It plays a catalytic role in this respect in Poland.

There are two other initiatives which could be treated as a bilateral debt-for-environment swap mechanism. These are:

- Convert 10% of the Polish debt with Finland (US$17 million) by support-
ing the purchase of Finnish goods and services by Polish entrepreneurs
(The Polish–Finnish Task Force Agreement, 1990).
- Convert 50% of German 'jumbo credits' from the 1970s for several aims
including the environmental sector. The money is managed by the Polish
German Foundation. For the last four years about US$15 million have
been spent on environmental projects.

An innovative financial instrument which serves environmental protection is
the Bank for Environmental Protection (BEP) which was established in 1992.
This bank is a unique solution in this field throughout the world. It gives both
commercial and preferential loans for investments in the environmental protection sector. The National Fund for Environmental Protection and Water
Management is the main shareholder of this bank and covers the difference
between commercial and preferential loans. Recently the importance of BEP
has been growing in the banking sector. It is now considered as one of the
strongest private banks in Poland.

The activity of international financial institutions in the environmental
protection area during the last five years also should be mentioned. The
European Bank for Reconstruction and Development has had no success in
this field so far. In contrast, the World Bank has signed several useful
agreements with the Polish Government concerning loans that serve, directly
or indirectly, environmental protection in the country. The main loans are the
following:

- Reform of environmental management methods (1990, US$18 million).
- Development of the forestry management (1993, US$146 million).
- Improvement of the efficiency of urban heating systems (1992, US$340
million).
- Support for Polish companies that carry out geological exploration for
natural gas (1992, US$250 million).

In conclusion it should be stressed that due to several innovative financial
instruments on the global scale, a solid foundation for fast progress in the
improvement of environmental conditions was established in the last few
years. All of them are functioning very well. The Polish market (about US$1
billion yearly) becomes increasingly more attractive for foreign companies.
The ecological fair POLECO, organized in Poznań every year since 1990, is
the biggest such forum in the countries of Central and Eastern Europe. It is
quite realistic that the Polish market in the environmental sector will absorb
funding at the level of US$25–35 billion until 2010. This will allow Poland to
reach ecological standards that presently exist in the USA or in the countries
of Western Europe.

International co-operation

The international co-operation of Poland in the previous political system was highly ineffective and was concentrated only on the COMECON countries. Agreements, signed at this time, only had a declarative character and had no impact on domestic progress.

After the fall of the Iron Curtain, Poland, as other post-Communist countries, opened its borders for close international co-operation not only at the official (governmental) level but also at the regional and local levels and among entrepreneurs.

Attention is especially focused on contracts with the European Union due to the process of adaptation of Polish law and ecological standards to those of the European Union (EU). The PHARE Programme is one of the tools supporting this process.

Intensive contacts exist between Poland and the Organization for Economic Co-operation and Development (OECD), especially in the following areas:

- Control of the international trade of chemical substances.
- Environmental protection in the energy sector and measures for saving energy.
- Elaboration of new ecological standards for raw materials and products.
- Problems connected with implementation of sustainable development.

In 1994 Poland was chosen by the OECD as the first non-OECD country for the Environmental Performance Review. The conclusions of this report are very valuable in terms of raising the effectiveness of the country's work in the field of environmental protection.

Poland also has good contacts with several United Nations agencies such as the United Nations Environmental Programme (UNEP), the United Nations Development Programme (UNDP) and the Commission on Sustainable Development (UNCSD). UNDP Poland for 1992–96 concentrates on measures for development of ecological education and training of Polish specialists working in this field on the local level.

The relations with UNEP have been intense for the last five years, especially in regard to Poland's signing of new regional and global conventions. Poland prepared a Country Study concerning biodiversity in the framework of working out the convention in this field (Rio de Janeiro 1992). Presently the Country Study on an Inventory of Greenhouse Gases and Measures for Reducing their Emission is being prepared.

Poland received two grants from the Global Environmental Facility (GEF):

1. For establishment of a bank of forest genes (US$4.5 million).
2. For conversion of coal into gas for heat supply in urban systems (US$26 million).

As many as 11 Western countries support Polish efforts in the environ-mental protection area by giving technical and financial assistance. Hundreds of consulting firms are active throughout Poland. At the same time, Polish experts are trained both abroad and at home. This is one of the reasons for the quick progress made in Poland in improvement of the state of the environment. In addition to this, both government and foreign non-governmental organizations give comprehensive support for the Polish NGOs continually.

Co-operation with neighbouring countries is also of great importance for Polish foreign policy. In this respect, good contact with Germany has high priority. Since re-unification of this country, Polish–German co-operation in the environmental protection area is very close indeed. The Polish–German Council on Environmental Protection was established at the ministerial level in 1991. It is the appropriate forum for regular discussions about all issues connected with contacts between these two countries. In 1992 the Agreement on Co-operation on the Frontier Waters as well as the General Agreement on Co-operation between Poland and Germany were signed.

The basic agreement between Poland, Germany, the Czech Republic and the European Union concerning the purification of the 'Black Triangle', which was signed in 1991, is very important. Since 1991 real progress in decreasing acid rain in this region has been noted. The Secretariat of the Black Triangle Commission is located in Usti nad Labem (Czech Republic).

Poland, together with the Slovak Republic and the Ukraine, has undertaken the very interesting initiative of creating the International Biosphere Reserve in eastern Carpata. This Reserve is situated in a triangle between these three countries. This mountainous region still retains its virgin beauty and bio-diversity. The site recently was added to the UNESCO List of Global Natural Heritage Areas.

A substantial part of international co-operation in the environmental pro-tection sector is based on international treaties and conventions. They are voluntary, but altogether they create international ecological law which stimulates pro-ecological activities. Poland has signed about 40 conventions. The most important are the following:

- Helsinki Convention on the Protection of the Marine Environment of the Baltic Sea Area (1974, new version 1992).
- Geneva Convention on Long-Range Trans-Boundary Air Pollution (1979).
- Ramsar Convention on Wetlands of International Importance (1971).
- Vienna Convention on Protection of the Ozone Layer (1985) with Montreal Protocol (1987) and London Amendment to the Protocol (1990).
- Basel Convention on Control of Trans-Boundary Movements of Hazardous Wastes and Their Disposal (1989).

- Helsinki Convention on Protection and Use of Trans-Boundary Water-courses and International Lakes (1992).
- Geneva Convention on Trans-Boundary Effects of Industrial Accidents (1992).

Moreover Poland ratified two conventions of great importance for the future of the world community:

- Rio de Janeiro Convention on Biological Diversity (1992).
- New York Framework Convention on Climate Change (1992).

Poland is very active on the international scene. This engagement is an additional stimulus for fast progress in the purification of its environment and effective preservation of the country's natural beauty.

6.4 PARTICIPATION OF SOCIETY IN ENVIRONMENTAL PROTECTION

Poland has a long tradition in social actions for nature protection. The Polish School of Ecologists was famous in the world before the Second World War. The first independent environmental movement – the Nature Protection League – was created in 1928. During the Communist regime, there were no conditions for independent activities until the rise of the 'Solidarność' Trade Union in 1980. The first post-war organizations of ecologists were created just then. The Polish Ecological Club was established in 1980 among others. It was, and still is, the biggest and the most serious non-governmental Polish organization in this field. Unfortunately, martial law closed this first period. Nevertheless by the end of the 1980s, the new wave of informal environmental groups had taken place. They were created mainly at the local level. Their existence was one of the reasons the Communist Government made the first reforms in the environmental protection sector.

In 1989, during the historic 'round table' negotiations between the Government and the opposition, environmental problems were included as one of the major topics. The opposition made the proposal that a programme be created which would consist of a set of measures for reversing the unfavourable trend of environmental damage in the country. Many good concepts from this programme were included in the State Environmental Policy adopted two years later – in May 1991 – by Parliament. Presently, this is a basic document for concerted and effective actions at all levels in Poland.

From 1989 to 1991, many new pro-ecological groups and organizations were established. About 200 such groups declared in their statutes that environmental protection and ecological education were two of their main tasks. Several 'green parties' were founded and started competing with each other. In 1991 the Ecological 'Fraction' of the biggest post-Solidarity party –

the Democratic Union – was created. The biggest, visible success of these organizations took place when, as a result of their actions, the establishment of the first nuclear power plant in Poland (which was 40% complete) was stopped and attempts to build the second nuclear power plant also were defeated. The Polish ecological movement also had a great part in effective action against the establishment of the huge coke plant in Stonava (Czech Republic), located about 10 km from the Polish border. However, the attempt to stop construction of the dam in Czorsztyn in the Carpata Mountains was a failure. In this case the non-governmental opposition had no meritorious arguments for stopping the proposed project. In fact many young NGO members were frustrated and felt they were manipulated by leaders of the movement. This failure was the first sign that the importance of independent pro-ecological organizations on the political scene was diminishing.

At present (1995) there are 300–400 NGOs. The majority have been established often just for protection of one valuable object or local territory, and this gives them limited importance. Only a few of them are active at the national level. Although they issue many magazines and newsletters and organize many actions like 'Earth Day' or 'Cleaning up the Earth', the significance of this movement in society has lately decreased. Obviously, the improvement in the state of the environment is one of the reasons for this decline. Economic problems are more important and urgent for the society now than environmental issues. Local groups often appear *ad hoc* for co-ordination of public protests against certain investment projects, especially in the location of a new refuse dump or waste incineration plant. It is symptomatic that presently on the political scene the Green Party has no members in Parliament. The Green Movement in Poland has become increasingly weak in spite of financial and organizational support from foreign countries.

It is expected that in the near future, in a new phase of the pro-ecological movement in Poland, the number of organizations will decrease. Simultaneously the strongest of them should be more active in many constructive actions at the local and regional level, thus raising public awareness on environmental issues. Nowadays, this awareness is territorially diversified and rather superficial. There is no doubt that the highest ecological awareness is characteristic of the citizens of towns and industrial regions.

Recently, environmental consciousness has risen substantially in the most valuable regions which are distinguished from the nature protection point of view. There are tourism benefits connected with the preservation of such areas. Unfortunately, awareness of this is still superficial because it concerns rather general questions like maintenance of a clean environment (i.e. pre-selection of wastes in a kitchen, not throwing away litter in forests, parks or streets) rather than more concrete issues. All the time the awareness of the common Pole is still at a low level. Changes are very slow because they concern mentality and, therefore, this is a task which will take decades. As a result just the ecological education of the entire society, both formal (in schools) and informal (by mass-media), plays the crucial role as a long-term investment in the environmental protection sector.

At the moment, ecological education in Polish schools is either presented as a special course in primary schools or at technical universities, or it is included in such courses as biology, physics, chemistry and geography in the secondary and high schools. In all agricultural universities and teachers' colleges, the environmental protection course is included in the syllabus as a basic course compulsory for students of all faculties.

The presence of a topic concerning ecological education is not the most important element in the programme of teaching the youth. Even more important is the form of presentation. All traditional forms of teaching which present this subject as an additional part of knowledge are boring and ineffective. Teachers should have tools like video-tapes, attractive folders and kits for laboratory work as well as for field study. They should show that protection of the environment is part of the right human approach to life, that it is a precondition for the survival of human civilization. Such presentation of this subject is still very rare.

The informal ecological education of society is also far from being satis-factory. All actions organized by the Ministry of Environmental Protection such as training courses, contests, sponsoring of mass-media programmes or organizing such nationwide actions like 'Earth Day' (in April) or 'the Day of Environmental Protection' (in June) have a very limited influence on society. Expenditure in this field from 1991 to 1993, budgeted from the Ministry of Environmental Protection and the National Fund for Environmental Protection, amounted to only about US$6 million. This is too little to make any progress in as important a matter as changing the approach to the environment in a nation of 40 million people.

Especially urgent seems to be the way ecological themes are presented in the most powerful medium – television. Presently such programmes are trans-mitted in the less attractive hours and often in a boring form. In contrast, the Western-style model of consumption is often advertised in TV in the form of aggressive, colourful, attractive advertisements and films. Such shaping of the society's consumption patterns is against the idea of sustainable development declared by the Polish delegation on all international forums for the last few years. Polish society has preserved many habits valuable from the perspective of sustainable development, but nowadays Western-style consumerism is rapidly taking place in the whole country. It is dangerous to neglect shaping the right consumption patterns in Polish society. Poland could become a country with good environmental conditions but with a low standard of life and quickly lose the entire richness of its traditional and cultural regional diversity. For this reason, effective and comprehensive ecological education of the whole society is of basic importance for the future of the Polish nation. There is still much to do in this most important field.

6.5 CONCLUSIONS

After five years of deep political and economic reforms, it is clear that the unfavourable trends of environmental damage have been reversed. Very innovative mechanisms, established at this time, are a good basis for further quick progress in this field. Nevertheless, many problems are still waiting for resolution while some other new problems are already developing and becoming more and more serious:

1. *Air protection* Despite big progress in the reduction of emission of SO_2, NO_x and particulates, Poland is still among the most polluted of the European countries. The problem of pollution generated by the transport sector becomes increasingly serious. Without drastic controls, it will be a main source of air pollution in the future. Actions undertaken so far for stopping this trend have turned out to be ineffective. Poland does not have any national long-range programme for development of the transport sector in cities and for long-distance connections such as exist in the rural regions.

2. *Water protection* The progress in this sector is especially spectacular. It is expected that in the next 15–20 years, all sewage will be purified in a proper way. At the same time the shortage of drinkable water for citizens of many towns will become a major problem. The expected man-made climate changes will make this problem more serious. The problem of desalination of mine waters is also very difficult to solve both from the technological and from the financial point of view. It has become the main cause of pollution in two of the biggest Polish rivers – the Vistula and Odra.

3. *Waste management* No substantial progress has been observed in the last five years. Just now there are some signs of change in this situation. Safe utilization or neutralization of hazardous waste as well as liquidation of thousands of illegal refusal dumps in the countryside are definitely the crucial problem. In Upper Silesia, it is necessary to put all waste from coal washing in mines together with fly ash from power plants which are located in this most industrialized region of Poland.

4. *Nature protection* The main task is to create new protected areas in the forms of national parks, biosphere reserves, and landscape parks. The promotion of pro-ecological agriculture and ecotourism in the most valuable terrains close to Nature Protected Areas also should be treated as a very important issue. Current activity in this area is so far inefficient.
   Pro-ecological forest management also should be promoted on the large scale. The pro-ecological function of forests is presently more important than the production of timber. Forests are especially rich ecosystems and play a crucial role in the preservation of biodiversity, in maintenance of appropriate groundwater levels and in absorbing CO_2. It should not be forgotten that forests are also an effective natural measure against floods

and soil erosion. The intensive education of foresters in such a way of thinking should have more attention.

5. *Ecological education*   This seems to be the most neglected area. Present expenditure for this purpose is insufficient. Public awareness in the pro-ecological approach is still weak and superficial. Even among the members of the Cabinet the opinion dominates that environmental protection, and even sustainable development of the country, is solely the duty of the Ministry of Environmental Protection. A similar way of thinking is presented by the majority of Parliament members and other influential persons like journalists, entrepreneurs, teachers, and priests. Intensive ecological education should be done first of all in these very influential circles. Only with the participation and full engagement of these groups can ecological awareness of the entire society be raised. Unfortunately, so far there is no national strategy for the next 15–20 years and no co-ordinated action at the national level.

6. *Participation of society in the environmental protection decision-making process*   Not much of this occurs at the local level. Decisions still are undertaken only by clerks while public protests are ineffective as they are often delayed. At present, any procedure of negotiations between investor, local authorities and representatives of society does not exist. This question should be taken into consideration more seriously in the near future.

7. *Eco-labelling of products*   This turns out to be a good measure for raising public ecological awareness, but so far such a system does not exist in Poland. The movement of independent consumer organizations is still in an embryonic phase. In the future they should play an important role in the goods market indicating which products are ecologically sound and which are dangerous for the environment or consume too much energy.

8. *Sustainable development*   All these problems should be part of the sustainable development comprehensive strategy of Poland for the next 20–30 years. This should be a complex vision of development for the whole country with descriptions of mechanisms for its achievement. The strategy should describe the future models of industry, farming, forest management, transportation, services, housing, health care, tourism, culture and education. It is obvious that environmental protection would be included in the majority of these sectors because only in this way would it be really effective, not just for the next four years but for the twenty-first century.

REFERENCES AND FURTHER READING

Andrzejewski, R. and Weigle, A. (1992) *Polskie Studium Roznorodnosci Biologicznej* (*Study on Biodiversity in Poland*). UNEP.
Głowny Urzad Statystyczny (Main Statistical Office) (1991–95) *Environmental Protection in Poland*.

Kassenberg, A. et al. (1991) *Outline of a Regional Policy for the Eco-Region Green Lungs of Poland.* Bialystok.

Ministry of Environmental Protection, Natural Resources and Forestry (1990) *National Environmental Policy of Poland.* Warsaw.

Ministry of Environmental Protection, Natural Resources and Forestry (1991) *The State of Environment in Poland – Damage and Remedy.* Warsaw.

Ministry of Trade and Industry (1994) *Energy Policy of Poland Till 2010.* Warsaw.

Nowicki, M. (1993) *Environment in Poland – Issues and Solutions.* Kluwer Academic Publishers.

Nowicki, M. (1995a) *Debt-for-Nature-Swap.* Conference on Economic Instruments for Sustainable Development, Prague/Pruhonice, January.

Nowicki, M. (1995b) *Poland's Experience of Financing Environmental Investments.* Workshop on International Financial Instruments in CEE Countries, IIASA, Laxenburg, March.

OECD (1995) *Environmental Performance Review – Poland.* Centre for Co-operation with the Economies in Transition, Paris.

State Inspectorate of Environmental Protection (1995) *The State of the Environment in the Katowice Voivodship in 1994.* Katowice.

CHAPTER 7

Slovak Republic

Mikuláš Huba

7.1 BACKGROUND

Brief characterization of Slovakia's environment

Several facts, immediately evident with a glance at a map, determine from the environmental point of view the character of Slovakia, the eastern part of the former Czech and Slovak Federal Republic. Slovakia covers an area of almost 50 000 km^2 and has a population of over 5.3 million inhabitants.

- The country is situated in the absolute centre of Europe. It is on the borderline between the cold north and the warm south, but also between the ocean influenced oceanic west and continental east. This is one of the preconditions of the country's varied and unique natural, cultural and scenic character.
- The country is situated in the principal European watershed (the so-called 'Roof of Europe').
- The essential part of its rivers flows into the Black Sea, the rest into the Baltic Sea. The rivers' wateriness, with the exception of the Danube River, is relatively small.
- The southern part of the country is lowland with a flat and hilly character (40% of Slovakia's territory). Its central and northern part is highland with a mountainous character (60% of Slovakia's territory).
- The landscape has the chessboard structure of the Carpathian mountain system which can be characterized as an alternation of mountain ranges and inter-mountain basins. The mosaic-like character is intensified by the geologically varied nature of the mountain ranges.
- The spatial distribution of forest and non-forest soils (land use in general) corresponds almost consistently to the existing 'antagonism' between the ranges and basins. In other words, mountains are almost totally forested,

The Environmental Challenge for Central European Economies in Transition.
Edited by J. Klarer and B. Moldan. © 1997 John Wiley & Sons Ltd.

while lowlands and basins (with the exception of the Záhorie lowland in the western part of Slovakia which has mainly poor sand soils) are almost totally deforested.

- The Western Carpathian vault rises towards the centre and declines towards the edges. It is dominated by the High, the West, the Belianske and the Low Tatras Mountains. Most of the landscape above the upper forest border is linked to the mountain ranges.
- The Danubian lowlands have very fertile soils. They are predominantly chernozems and brown soil types.
- Most settlements and productive capacities are situated in relatively limited bands of river terraces along the most significant Slovak rivers.
- Inter-mountain basins are separated from each other by the mountain barriers and are interrupted only by narrow valleys or mountain saddles.
- Most land suitable for agricultural use has been transformed into arable land.
- The fact that Slovakia is situated close to regions with considerably polluted atmospheres intensifies the deposition of harmful pollution on the territory of Slovakia.

The Socialist revolution after the Second World War also represented a revolutionary impact on the character of the Slovak environment. How could we characterize this briefly? Well, as an inappropriate application of foreign, i.e. Soviet, models and methods to the Slovak environment. Huge 'Soviet like' constructions and the priority given to the development of the most harmful heavy industrial branches were not suited for the previous character of Slovakia, its rich micro structure, great natural and cultural heritage, healthy and beautiful environment so typical for this 'gem in the heart of Europe'. In the same way, the application of Soviet kolkhoz methods and large-scale ploughing of the virgin land did away with the values of the beautiful Slovak cultivated rural landscape.

Legislative, conceptual, organizational and technical ignorance of the scope of problems of communal waste caused the Slovak countryside to be impaired by thousands of elementary tippings or rubbish heaps. The centralization and concentration of people and activities, in contradiction to Slovakia's natural character and traditions, led to overburdened, even collapsing structures and large squalid areas. Heavy industry became fatal for Slovakia for several reasons. It naturally conflicted with the needs and character of a small country with a dissected relief which was not very rich in raw material and power sources. Heavy industry often had an absurd character and localization. Chemical works and colour metallurgy plants were localized in un-aired, narrow mountain valleys or basins. Since almost no attention was paid to the modernization of the unnecessarily megalomaniac plants in the past, we cannot be surprised at the bad health condition of Slovakia's population and its environment today. In addition, Slovak heavy industry was characterized by intense engineering, chemistry, magnesite processing, paper mills, and cement. Almost every town had its harmful factory.

Industrialized agriculture also caused the deterioration of the environment. Hygienic problems stemmed from mammoth large-scale breeding farms which were not adequately equipped with sewage systems, protective greenery and other important infrastructure. The unification of fields formed endless areas of mono-cultures without a single spot of greenery, without a single refuge for wildlife, without former terraces and other anti-erosive barriers. Agriculture was deprived of aesthetic qualities, life and ecological stability. The rate of chemical action was so high that, apart from other things, it actually had a serious impact on life in soil and underground waters.

Slovakia's forests were also affected. The trend of worsening conditions of Slovakia's forests is serious and is perhaps the most rapid in Europe. While 2.4% of the area covered by forests was recorded damaged by emissions in 1970, the rate was 8.5% in 1975, 15.6% in 1985, and more than 50% in 1990 (SZOPK, 1990).

Some environmental statistics and trends (MOE, 1993; Statistical Office of Slovak Republic, 1994/5)

In those parts of the major Slovakian rivers where monitoring actually takes place, there is severe pollution: 70–80% of the rivers belong to the two categories (fourth and fifth) of greatest pollution. Megalomaniac intentions to construct waterways endangered the last natural parts of the bigger Slovak rivers.

Biochemical consumption of oxygen in 1991 was 31% of the 1985 figure, and chemical consumption of oxygen was 24% less than in 1985. This decrease in pollution is mostly a result of the economic downturn. Whereas in 1970 187.3 million m^3 of wastewater were released into the public sewage system, in 1990 the figure was 492.7 million m^3. Groundwater quality began to deteriorate significantly. In 1991, 87% of water samples were judged unsuitable for consumption as compared to only 63% in 1983. The specific consumption of drinking water in 1991 – 408 litres per capita daily – was almost double that of Austria.

Slovakia remains ninth among European states in sulphur dioxide emissions and produces four times more SO_2 emissions than neighbouring Austria which is about two times greater with respect to population and territory. SO_2 emissions amounted to 120.4 kg per capita in 1985 while in 1991 the rate was at 'only' 83.0 kg per capita. Large transboundary SO_2 emissions come from Poland, the Czech Republic and Hungary. CO emissions, which in 1985 reached 65.7 kg per capita, are declining (to 5.4 kg per capita in 1991), but emissions of nitrous oxides increased from 38.2 kg per capita in 1985 to 44.2 kg per capita in 1991. Solid emissions exceeded 226 000 tonnes in 1991. All these data include domestic stationary sources only. Transportation, transboundary emissions, and household heating data are excluded.

Waste poses great problems for the environment. Of the 34 million total tonnes of waste produced in 1991, 3.3 million tonnes were hazardous waste. There are 7204 recorded landfills, only 335 of them licensed. Radioactive waste from nuclear facilities is currently stored in nuclear power plants where the storage capacity in the optimal case will be reached by 1997. The aggregate production of radioactive waste by the year 2030 is estimated to be 38 000 m^3 liquid and 20 000 m^3 solid waste. In addition, there probably will be other wastes from the damaged nuclear power plant A-1 in Jaslovské Bohunice.

The overall per capita consumption of primary energy is 3–7 times greater than in developed countries. In 1989 the Slovak Republic produced 3.5 million tonnes of pig iron and 4.7 million tonnes of steel. However, there is a shortage of domestic sources of high-quality raw materials (excluding non-ore material and magnesite). These have been intensively exploited regardless of the local environmental situation. The result of this production orientation is that industry consumed 63% of electricity in 1991. Per capita consumption of electricity increased by 33.4% between 1980 and 1990.

Infrastructure is overloaded by inefficient transport of loads. Slovakia, due to its Central European location, seems to be favourable for transit routes. On the other hand, the mountainous character of the country, heavy population density in valleys and basins, and overburdening of the environment are serious barriers to such tendencies. The Slovak Government is fighting for the opportunity to build a trans-European highway with a north–south connection. Another goal of the Government is to build new waterways with international connections. This would cause many negative environmental impacts through the March and Vah Rivers.

Destruction of historical centres of towns and villages is the result of centralistic tendencies related to ideas of revolutionary urbanization and industrialization, as well as the anti-historical ideology of the so-called socialist, all unified modernism. Town and/or territorial planning has been highly centralized. The creativity of architects and urbanists has been subordinated to the dictatorship of huge state prefabrications and concrete producing firms.

An emerging new problem is the development of a market of environmentally harmful products such as cigarettes, cars, and non-recyclable packing materials along with aggressive promotion and advertising of such products. The promotion of tobacco products in the capital of Bratislava has caused public indignation.

The state of the environment plays an important role in the sphere of the population's morbidity and mortality. The average life expectancy at birth for men (66.5 years) is six to seven years less than in developed countries. For women the life expectancy of 75.3 years is three to six years less. Mortality, which increased by 0.2% from 1960 to 1980, has since stagnated. More than one-half of the diseases are heart and vascular (53.3%), one-fifth are malignant tumours (19.0%). The morbidity and mortality from malignant tumours has nearly doubled as compared with 1965.

Transboundary Problems

Air pollution Slovakia predominantly receives north-west winds. This results in the transportation of emissions from Upper Silesia, Krakow and the Katowice regions. This transboundary air pollution has a destructive impact on the borderline mountain forests and sensitive ecosystems above the upper forest lines – areas that mostly are national parks or protected landscape areas.

Water pollution Slovakia is located in the main European watershed. The country breaks down into the Danube and Vistula systems. The river network of Slovakia is 'feather-shaped', created by one central river supplemented by many relatively short tributaries. Slovak rivers (with the exception of the Danube River) are small and receive water pollution from heavy polluting industry, agriculture, settlements, transport and other human activities. Water pollution from Slovakia is carried by the Danube, Hornád, Bodrog and Poprad Rivers. Water pollution is carried to Slovakia through the Danube, March, and Uh Rivers.

Gabčíkovo Dam The Gabčíkovo–Nagymaros Dam System Project (Gabčíkovo), is located at the Slovak–Hungarian boundary. It has become undoubtedly one of the best-known environmental–political problems in Central Europe. Many politicians try to present Gabčíkovo as a conflict between Slovaks and Hungarians. A much better interpretation is that Gabčíkovo is mainly a conflict of two opinion groups, the members of which are on both sides of the Danube River. This is actually a conflict between the Slovak and Hungarian representatives of the old-fashioned technocratic approach to nature and the Slovaks and Hungarians who support preservation of nature and cultural values and sustainable development of this region. According to Šíbl (1993), environmentalists, scientists and common people on both sides of the Danube River have fought for more than 15 years against this monstrous project. Hungarians, Slovaks, Czechs, Austrians and people from other nations have joined together in the biggest environmental battle in the history of the region. On 23 October 1992, the Hungarian Government turned to the International Court in the Hague requesting it to decide on the Gabčíkovo case.

Nuclear power plants In Slovakia, one nuclear power plant is in operation, V-1 and V-2 Jaslovské Bohunice. A-1 Jaslovské Bohunice has been closed since 1976. The other plant is under construction in Mochovce. All these nuclear power plants are of the Soviet type of construction and are located approximately 100 km away from the borderline of nuclear-free Austria. The financial involvement of EBRD in the Mochovce project is a permanent matter of discussions in the European Parliament and relevant EU Commissions. The campaign of environmentalists against the completion of the Mochovce Nuclear Power Plant has created the greatest 'green' activities held recently in Europe.

Hazardous and toxic wastes   Especially dangerous are tendencies of Western-based companies to export hazardous and toxic waste to Slovakia in spite of international agreements, especially the Basel Convention, which require strong regulation of such activities.

Impact of political and economic transformation

Political and Economic Determination

Two major political events in the Slovak Republic, the 'Velvet Revolution' in 1989 and the 'Velvet Divorce' in 1993, had, and continue to have, major impacts on environmental quality, policy and the development and growth of domestic markets, technologies and services. The 'Velvet Revolution' stimulated the development of environmental markets by improving environmental regulation and enforcement. On the other hand, the economic transformation significantly depressed certain sectors of the economy, also reducing available sources of environmental investment. The 'Velvet Divorce', and the political instability that followed, paralysed economic restructuring and the development of the private sector leaving only the 'fragile' government and self-governments as sources of environmental investment (Macek and Urbánek, 1994).

Recent Development of the Economy with Emphasis on Macro-economics and System Measures

Recent economic development shows that desirable structural changes which would aim at the growth and preference of environmentally less harmful industries are largely lacking. In addition, no consistent structural changes in the spheres of technological and infrastructural development, which would ensure the reduction of energetic, material and raw material demands, are realized. Such measures would result in the positive development of the state of the environment. From the point of view of the industrial composition of production, no distinct growth of new technologies orientated to improvement of the environment (sewage water cleaning stations, waste recycling and separation lines, and equipment for alternative energy) has occurred. This sort of industry is generally considered as structurally very positive with good prospects for outlets and exports.

With regard to legislation and the economic system, there does not exist yet real support and promotion for pro-environmental enterprise, trade, consumption and behaviour. The internalization of external environmental costs into production and product costs, though generally recommended, was not realized. In other words, taxes, charges or fees do not reflect ecologically harmful production and products. This, consequently, results in the fact that production methods and products harmful to the environment are indirectly favoured as compared to environmentally-friendly ones.

Due to the significant influence of powerful industrial lobbies, not much is contributed by the greatest polluters. As a matter of fact, in environmentally harmful heavy industries like heavy chemistry, pulp, pig iron and crude steel production, and aluminium, the consumption of energy dropped more slowly than the general production volume. Production of goods dropped from Sk 624 840 million in 1989 to Sk 317 259 million (Slovak koruna) in comparable prices in 1993 (US$1 = 30 Slovak koruna). At the same time (1989–92), the final consumption of fuels and energy in Slovakia dropped in the industry and construction sectors only from 370 027 TJ (terajoules) on 9 November 1989 to 314 990 TJ in 1992 (*Statistical Yearbook of the Slovak Republic*, 1993/4).

Hesitation in the privatization process, on the one hand, and an insufficient consideration of environmental aspects within this process on the other, as well as delays in efficient functioning of the Act on Bankruptcy (which should consider not only the direct financial, but also the environmental debt), led to an artificially prolonged life of polluting, inefficient, outdated production. This has prevented the beginning of modern and environmentally-friendly production.

There is an overall opinion that the protection of the environment and the transition towards environmentally-friendly technologies and products represents a load or a delaying factor that the contemporary economy is not able to bear. In contradiction to that opinion is the recognized reverse truth, namely that these act as a real developmental stimulus providing a qualitatively new, sustainable, self-supporting and self-regulative type of development.

In the sphere of re-privatization of forests, no success in preventing the negative effects of this process was achieved. Common good interests are insufficiently protected. In some parts of Slovakia, there exists something like anarchy by some of the new owners, the heirs of the previous owners and now the new users of the forests.

It is absolutely necessary to prepare a solid code of conduct for Western investments in Slovakia and other countries which are similarly economically weak and inexperienced. Western investments sometimes involve transfer of outdated technologies, information espionage, and sell-out of the natural resources. It is necessary to make the process of foreign investment more transparent, to address priorities which will take into consideration the historically tested needs of this region, and at the same time use the positive experiences of economically more developed countries.

Budgetary, Domestic and Foreign Resources; Investments

In the last three years, the Ministry of Environment was the Ministry that received the least money from the state budget. In 1993, in comparison to 1992, only about 60% of the means was allocated to it while the state subsidy of the State Environmental Fund (SEF) dropped steadily:

– 1992: Kčs 950 million (ca. US$ 38 million, Kčs was the currency before the split of CSFR in 1992)
– 1993: Sk 440 million (ca. US$ 15 million)
– 1994: Sk 300 million (ca. US$ 10 million)
– 1995: Sk 250 million (ca. US$ 8 million).

This does not account for the effect of devaluation and inflation. Additional substantial income to the SEF is realized through the collection of various environmental fees and fines. However, the enforcement of these fees and fines is undermined by benevolence towards the polluters. Apologies are often made for the indebtedness and insolvency of such companies. The scope of activities financed by the SEF has dropped to a minimum.

State financial support for non-governmental environmental organizations dropped to about one-tenth compared with the year 1992. In the case of the Slovak Union of Nature and Landscape Protectors, for instance, this resulted in the end of the existence of an all-Slovakian professional staff structure.

Assistance from foreign sources used to be not transparent enough in the criteria and priorities for granting such assistance, as well as the adequacy and efficiency of granted projects. NGOs are not invited by the State to participate in the use of some relevant assistance funds, although their participation in the use of these resources is assumed by the donor. The orientation of international financing institutions (with the exception of GEF support by the World Bank) is often not supportive of environmental projects. In spite of its General Rules, the EBRD is supporting projects that are negative or problematic from the viewpoint of the environment.

On the other hand, we can welcome activities of some internationally based foundations that support NGOs in Slovakia. The Environmental Partnership for Central Europe, the Regional Environmental Centre for Central and Eastern Europe, and the Global Environmental Facility Foundation are addressing the needs of environmentally-orientated NGOs. Others such as the Foundation for a Civil Society or PHARE Foundation for Support of Civil Activities are orientated to all of the NGO sector, including the environmental part. In general, it can be stated that foreign assistance to the NGO sector is much more useful and effective than governmental help.

The lack of financial means provokes an exploitation of future nature resources. It means the risk of over-exploitation and/or export of inland resources, e.g. wood and non-renewable raw materials.

Practically all 'post-revolutionary' governments supported numerous environmentally-problematic projects without assessing environmentally friendlier alternatives. Examples are: the Gabčíkovo-Nagymaros Dam Project, the Mochovce Nuclear Power Plant, highway construction, car production, the reconstruction and expansion of the Žiar nad Hronom Aluminium Plant, the construction of water reservoirs, and plans to organize the Winter Olympic Games in Slovakia's national parks.

The liquidation of the crushed down nuclear power station A1 in Jaslovské Bohunice, as well as an end to the operation of the outdated V1, has been delayed indefinitely. In constructions like the Mochovce Nuclear Power Plant, capacities and capital are inert and missing in spite of the fact that more sophisticated investments in the spirit of sustainable development exist. Available finances, research potential, building and machinery capacities, foreign credits, as well as state guarantees, are blocked.

The construction of water treatment plants is delayed. This is a result of low priorities given to environmental investments. Subsidies of the State Environmental Fund, the main investor especially in the field of communal waste treatment, have dropped because of the poverty of town and village budgets, and insolvent industrial polluters.

Prices for public transport have increased. Now they are approximately five to seven times more than than they were from 1990 to 1995. In comparison, petrol prices have only increased twice in the same period. Car prices have increased only three times. If you buy a new car, you receive a tax reduction from your annual income. Service by train and bus lines has been reduced, as have connections. All of these actions encourage individual car transport which is more environmentally harmful.

Lacking Capacity to Tackle the Problems

Uncritical prioritization of economic transformation goals, postponement of the 'environmental transformation' and subordination of ecological values to economic ones prevails. This was especially the case from 1990 to 1992. The lack of competent, experienced and non-compromised persons in the post-revolutionary political and economical structures at all political and economic levels continues to be a serious issue.

A very dangerous situation both for the economy and the environment continues. There are indications that the old interest groups of the previous system are interwoven with the new ones arising in the country – the international lobbies or mafias. These groups aim at further parasitism, the ruin of the economy, natural resources, and cultural heritage. This affects the export of raw materials, the import of wastes and obsolete technologies and bad credits.

In Slovakia the tradition of dangerous personal unions, an issue which is not accepted in Western democracies, has been retained and even strengthened. A serious law regulating 'conflicts of interest' still does not exist. This also has negative impacts in the sphere of the environment (e.g. the Vice President of the Slovak Parliament from 1990 to 1992 was at the same time the director of the huge constructing firm which was building the Gabčíkovo Dam).

There is a lack of a 'sense for law', especially in the underestimated field of the environment, particularly in the area of nature conservation. For instance, the aspiration of Slovakia to conduct the Winter Olympic Games 2002 in the High and Low Tatras was supported by practically all 'top-level' politicians regardless of the fact that the organization of such a huge event on the territory of the national parks is not compatible with existing Slovak legislation, including the Constitution of the Slovak Republic.

The process of adoption of new environmental legislation is progressing with great difficulties – not only due to the lack of competent legislators who would be able to solve these wide and complex problems, but also due to the resistance of those in the economic sphere who often regard environmental protection as an undesirable barrier to realization of their plans. From 1990 to 1992 tensions and conflicts between the federal and the national levels existed. As a result the assignment of permanent responsibility was set back. This complicated the process of adoption and/or creation of environmental legislation.

The Slovak Commission for the Environment (SCE) was created mainly by people who dealt with the environment in different central bodies during the past regime. There is a lack of enthusiasm, a spirit of bureaucracy and ignorance with no significant exceptions. Among the positive activities of the SCE (which was transformed into the Ministry of Environment later) is the fact that in autumn 1990, it submitted to the Slovak Parliament a proposed law on the establishment of a separate, regional (local) environmental state administration subordinated only to the law and the SCE.

A general problem in state environmental protection, especially at the top level, is an obvious lack of people who are educated to combine protective education with protective conviction. After the 'Velvet Revolution', the situation with respect to this issue improved only slightly. Environmentally-orientated departments of universities produce yearly only an insignificant fraction of the total number of graduates. There are not enough really skilled lecturers who can combine scientific and pedagogical professionalism with an eco-philosophical background. Yet a greater handicap in the educational system is the absence of holistic, synthesis-orientated, and landscape, ecological or geographical approaches in the educational plans of all school types and grades.

Although the Slovakian environmental movement was versatile, inventive, altruistic, active and had revolutionary influence, it was also, quite logically, the movement which consisted mostly of enthusiasts and non-professionals in the sphere of environmental protection. In contrast, the Czech Republic, had a long tradition not only of formal scientific bodies but also of non-governmental ones. For example, the Ecological Section of the Biological Society of the Czechoslovak (actually Czech) Academy of Sciences consisted before the Revolution of about 500 environmental experts. In Slovakia there is obviously a lack of a sufficient amount of independent, ecologically-orientated specialists who could constitute a co-ordinated front. With all respect, it is not possible to build overnight a highly specialized, central, conceptual and decision-making body only with enthusiasts.

7.2 INSTITUTIONS AND POLICIES

Brief outline of the institutional set-up for environmental protection

Until November 1989, the responsibility for environmental issues was dispersed among several institutions, and no central body in charge of environment existed. Through the so-called 'guarantee system', different ministries were responsible for those aspects of the environment with which they were economically or administratively connected. Environmental NGOs several times before the 'Velvet Revolution' proposed the establishment of a Ministry of Environment. This was also one of the demands of the well-known eco-dissidents' publication, *Bratislava/aloud*. The idea to establish a central body for the environment was stated by the Slovak Environmental Forum at the beginning of December 1989. The Forum stated that it was necessary to create a cross-sectorial, or over-sectorial body, which would be headed by the Vice-Prime Minister on both the federal and national (Czech and Slovak) levels.

The primary idea was to integrate into a central body, which had no direct economic responsibilities, needed legislation, strategic, conceptual, co-ordinational and controlling responsibilities for a wide range of environmental issues, including land and forest protection, hygiene, and historical monuments' protection. The idea was based on the concept of a division of power between different spheres which would be independent of each other. The principle of environmental complexity also was part of this idea.

The result was (is) something like a compromise between these two pure ideas. A central body for the state environmental policy, the Slovak Commission for Environment, was established on 1 April 1990 – one-quarter of a year after the Ministry of the Environment of the Czech Republic was established. In the first period, the Commission was chaired by the Deputy Prime Minister. A new government was established after general elections in June 1990, and the Commission then was chaired only by the Minister of

Environment. After the general elections in June 1992, the Slovak Commission for Environment was changed to the Ministry of Environment.

At the federal level, the Federal Committee for Environment (FCE) was established on 19 June 1990. It mainly addressed general legislation, environmental policy and conception, and international aspects of environmental protection. The idea was not to create another ministry but a collective body composed of ministers of the environment, deputy ministers of foreign affairs, finance and economy and also chairmen of environmental committees of the three parliaments. The responsibilities of the FCE were reduced rapidly just after the elections in 1992, and it disappeared with the split of the Federation in 1993.

Just after the 'Velvet Revolution' in November 1989, parliamentarian environmental committees of both the Federal Assembly and National Councils also were established. After the split of Czechoslovakia and the changes of the structure of Slovak Ministries and other state central bodies, the Ministry of the Environment assumed the responsibilities of the former Federal Committee for Environment and the Slovak Geological Office.

The original responsibilities of the Slovak Commission for Environment (SCE) corresponded to those of the Ministry of Environment in the Czech Republic in the following spheres: air protection, nature protection, territorial planning and building laws, installing of the information system about the environment, the main state control of environmental matters, water quality/quantity protection and conceptional questions of waste management. The Czech Ministry was responsible for the following additional spheres: agricultural and forest land protection, geological research, raw material protection, and ecological supervision over mining. The SCE was paradoxically orientated towards gaining responsibilities from the Federal Committee of the Environment (FCE), although its position and responsibilities within the Slovak Government were relatively weak ones. Its responsibilities were enlarged by incorporation of the Slovak Geological Office.

Slovakia succeeded in the creation of a two-level system of specialized state district/sub-district administration. Since January 1991, 38 district and 121 sub-district (county) environmental offices directly subordinated to the Ministry of the Environment have been working. The positive feature of such a system is that these environmental authorities do not depend on the general state administration but are directly subordinated to the Ministry of Environment (previously SCE). The district environmental authority is the appealing organ for the sub-district one. The imperfection of such a system in practical life (according to Zamkovský, 1992) arises from the fact that sub-district environmental authorities have a minimum of autonomy from the district ones. Sub-district environmental authorities do not have personnel sections, therefore they cannot control their own structure and wages. The head of the district environmental authority can appoint and remove the head of sub-district authority.

As the district environmental authorities are mainly conceptually orientated, an essential part of the work is executed by the sub-district authorities. The law enabled the SCE to concentrate on the execution of more district or sub-district authorities rather than a single one. This was implemented, for instance, in the sphere of water protection. The authority to give permission for water management activities was given to the regions which were the centres for river basin management authorities with their teams of water economists and laboratory equipment (approximately eight districts belong to one region).

The practical results achieved by the specialized state environmental administration could be described as very positive despite ongoing financial and technical problems. Despite the unambiguous necessity of posts being occupied by interdisciplinary experts, the present State demonstrates a preference for civil engineers and architects. However, initially some posts were occupied by environmental professionals who brought in an unofficial attitude and gradually gained authority. During 1991 and 1992 as new environmental legislation was adopted which dealt with the creation of the State Environmental Fund, the administration and procedures in air pollution control, as well as complete new legislation concerning waste management, responsibilities of the SCE were enlarged and justified. At the national level, there exist some other bodies supervised by the Ministry of Environment: the Slovak Environmental Inspectorate, the Slovak Environmental Protection Agency, and the State Environmental Fund.

Part of the environmental responsibilities belong to the local governments according to the respective 1990 law. In Slovakia, no district self-governments exist yet. The division of responsibilities between the state administration and self-governments is given both by the Law on State Administration for the Environment and the Law on Local Governments and is further determined by special environmental laws such as the Clean Air Act and the Waste Management Act. The responsibilities of self-governments are, for instance, the arrangement of local territorial plans. The local government authorities determine emission limits, fines and permissions for small-scale air pollution sources and regulate mobile air pollution sources. Local governments have large responsibilities in spheres such as waste management and the EIA participation process.

Key players

In the REC (1994) report on *Strategic Environmental Issues in CEE*, the following environmental coalitions were identified as playing key roles in environmental protection in the post-Communist countries:

- National and local governments and politicians/political parties
- Business sector and trade unions

- NGOs, scientists and media
- Consumers (the general public).

The first 'coalition' has been described previously. In a situation when the rule of law and the tradition of state environmental protection is not developed and stabilized enough, we can agree with the opinion, declared by the REC study, that the politician's and/or civil servant's perspective regarding environmental problems is strongly affected by his or her own role and function. Devising policies, implementing projects and enforcing the law is the role of the environmental administration. Within this body a formal, and sometimes bureaucratic, approach to emerging problems is typical. The position of the Ministry of Environment is rather difficult in opposition to the more powerful economic ministries in the Government. It does not have enough responsibilities for covering all sectors of the environment. However, the Government, Parliament, media and public are used to make the Ministry of Environment responsible for solving all environmental problems. According to Zamkovsky (1992), the Slovak Ministry of Environment is not flexible, co-operative and emphatic enough, and it is almost completely passive in the process of communication with the public. We can, however, mention two successes of the Slovak Ministry of Environment during its history so far: first, the preparation of *The Strategy, Principles and Priorities of the State Environmental Policy*; and, second, the creation and defence of the specialized district and sub-district environmental administration.

Non-environmental organizational spheres have a stronger influence on environmental issues than those that contain environmental terms in their name. We can talk about the following subjects:

1. Government and individual ministries (besides the Ministry of Environment).
2. District and local offices of state administration (besides the respective offices for environment).
3. City and communal self-governments (besides their departments of environment).
4. National Council of the Slovak Republic (besides the Committee for Environment and Nature Protection Preservation), and political parties and movements.
5. Economic sphere, economic sectors.
6. Public Prosecutor's Office, Courts, Supreme Control Office, etc.

1. The Government and Ministries

None of the governments after 1989, neither Slovak nor federal, succeeded in abandoning pre-November philosophy and practice related to environmental issues. This goes for the sphere of prevention as well as an old fashioned approach in the field of both system measures and individual projects. Economic policies, especially with respect to particular industrial projects and sites, were not changed. Governments (and parliaments) did not consistently make use of the fact that the whole system of legislation, including an entirely new tax system, was reconstructed. This could have represented one of the most efficient tools for the solution of environmental problems. A differentiated approach offering advantages to environmentally-friendly products and technologies and/or disfavouring the environmentally-harmful ones could have been provided.

None of the governments, in spite of the declarations of the right-wing ones, succeeded in forging system measures such as a systematic reduction of state subsidy for non-efficient and environmentally-unfriendly state enterprises. They were not able to stimulate efficiently the conversion of Slovak industry, did not have sufficient courage to bring through an Act on bankruptcy, or make it obligatory for environmental audits to be performed in the first round of privatization, or require compulsory application of least cost studies.

All governments and a great majority of the political parties agreed on direct or silent support for anti-ecological projects like the Gabčíkovo Dam, the aluminium plant in Ziar nad Hronom, Nuclear Power Plant Mochovce, and the entry of Poprad-Tatry for organizing the Winter Olympic Games.

The traditionally strong position of the Ministries responsible for industries and other economic sectors was not shattered even by the 'Velvet Revolution' or the following developments. With the exception of top party personalities at prominent posts of departments, the majority of important posts are held more or less by the same people as before 1989. These people also have conserved a style of thinking and working, centralism, monopolism, etatisme, etc. They have preserved projects that originated before November 1989. The momentum in their implementation has enriched old friends and contacts by new possibilities brought about by the fall of the Iron Curtain. The conception that industries should serve national needs is still being prepared without alternatives by the departments that will be directly engaged in its subsequent realization.

For instance, the energy policy creates a barrier for a real energy market which gives strong preference to energy saving measures or the support of alternative energy sources. The entire energy sector is under too heavy control by the state monopoly. It is highly centralized. Energy prices are deformed by direct and/or indirect subsidies. On the other hand, the real possibility of the private sector acting in energy production through demonopolization and decentralization is strongly limited by many direct and/or indirect barriers. These consist of artificial prices, lack of credits and state guarantees for decentralized energy producers on the basis of co-generation. There also is an absence of relevant economic and legislative tools.

Individual industrial sectors still exert strong influence upon the political establishment, executive power, and an important part of the media.

Environmental issues represent a traditionally undesirable burden for the Ministry of Finance. The Ministry of Interior still has not adapted to the loss of its powers to the Ministry of Environment in the area of regional state administration. The Ministry of Construction is attempting to shift a substantial part of power from the sphere of environment to its own sphere.

The result of the struggle over responsibilities is a preliminary victory for the MOE. Areas under dispute are connected with territorial planning, construction permit issues and state administration authority over responsibilities. The Ministry of Land Management, too, would like to get back water protection powers given to the Ministry of Environment in 1990. Economically orientated ministries and lobbies connected with the Government systematically ignore the possibilities to finish reconstruction of the powers of the Ministry of Environment in a complete form.

2. District and Local Environmental Offices of State Administration

With Act No. 595/1990 on State Administration for Environment, a substantial part of power to care for the environment was given to the district and local environmental offices. The rest remained in the sphere of influence of the Ministry of Interior and other central organs of State administration. The Act concerns the hygienic service, protection of forests, as well as the forest and agricultural soil, and care for monuments. The problem is that the Ministry of Environment only has responsibilities over a part of the environment. This means its authority in this field is partial only (in contradiction to its name and public perception). The remaining powers were given to the cities and local self-governments.

3. City and Communal Self-government

Since the creation of the new system of self-government in 1990, self-governments are fighting to receive some responsibilities from the state environmental administration. Some legal drafts concerning this change of responsibilities were prepared by the Government or relevant ministries in co-operation with the Association of Towns and Villages in Slovakia, but none of the proposals has been accepted by Parliament, yet. The position of the district and sub-district environmental administration is permanently threatened, not only by the interests of self-governments, but also by the general district/sub-district administration, as well as from the side of some ministries. There exists a permanent struggle over responsibilities, which originally (before the establishment of the Ministry of Environment and its regional administration) belonged to other ministries or regional authorities.

There are considerable differences between the various Slovak communes with respect to their environment, economic structure, population composition, and nature of the local educational institutions. Equally important is the staff working for the local self-government, especially the mayor, who according to the effective legislation possess relatively great powers. However, the possibilities of the towns and communes paying attention to communal environmental issues is limited not only by the shortage of finances, experience and imperfect legislation, but also by the lack of sound local patriotism which was almost eradicated by the previous regime.

As far as the rural settlements are concerned, in 1991 the Government of the Slovak Republic adopted a so-called Programme for Village Restoration. Unfortunately, due to the same reasons, this programme has been only partially, and at a minimum rate, implemented. Further perspectives for the environment lie in associations of the towns and communes (villages) on regional or common interest principles.

4. The National Council of the Slovak Republic, Political Parties and Movements

The parliamentarian system of political pluralism was formally re-established just after the 'Velvet Revolution'. A traditional spectrum of standard Western European parties does not exist yet. The majority of politically orientated members of the pre-revolutionary environmental movement was divided in principle into two parties (movements): the revolutionary movement Public against Violence and the Green Party. In 1991 the original Federal Green Party was divided into two parties. Today, both the former Public against Violence movement and the former Slovak branch of the Federal Green Party are parts of the Democratic Party. The second part of the former compact Green Party is called Green Party in Slovakia. Two of its MPs became members of the Slovak Parliament during the elections in 1994 as members of a left-orientated coalition called 'The Common Choice'.

None of the parliaments after November 1989 made consistent use of their powers in terms of lawmaking, initiative, and control mechanisms for improvement of environment. The number of deputies with environmental education or a priority interest in environmental issues is critically low. Since the 1990 elections this has steadily decreased (they even did not constitute a majority in the Committee for Environment, not to talk about other committees). A similar situation can be observed with political parties where environmental priorities declared in the 1990 election programmes of prac-tically all parties vanished or fell down in the priority lists.

For the 1994 elections, a number of environmental issues including strategies for sustainable development appeared in the programme of the Democratic Party and The Common Choice. The Green Party in Slovakia has two MPs and the Democratic Party only one representative in the Slovak Parliament. The Green Party is relatively active in environmental issues. The Democratic Party usually criticizes technocratic and state monopoly approaches (e.g. huge dam building and/or centralized energy production).

5. The Economic Sphere

The economic sphere still has not realized the inevitability of change of behaviour in an environmentally-friendly and efficiency-orientated society. In cases of state enterprises, there are some demoralizing direct and indirect subsidies – for instance in energy prices. The artificial low price of energy is one of the comparative advantages of the Slovak metallurgy, cement or heavy chemistry industries. The Slovak economy tries to compensate for the drop in consumption in home markets and the disintegration from the COMECON market by an effort to export commodities, even at dumping prices. Exports of raw material and semi-products cause further pressures on natural resources (wood, construction material, and metallurgical products).

In the agricultural sector the lower input of chemical fertilizers and pesticides, caused by the poor economic situation of the agricultural companies, has had a positive influence on the environmental situation. In terms of transportation development the lowering of transport load, due to the drop in transport of large-volume commodities, has been positive because it has caused a drop in the general impact of human activities on the environment. On the other hand, an environmentally-undesirable effect has been the development of the car and truck transport.

A rapid decrease of production volumes was recorded in construction. This resulted in a positive effect on the environment (with the exception of construction of sewage water cleaning stations, etc.). The diversity and overall quality of architecture and urban planning increased. More attention is being paid to the restoration of historical objects and intensification of the use of urban space. The effects of privatization and other features of economic transformation have been discussed earlier.

6. Public Prosecutor's Office, Courts and Other Relevant Subjects

In spite of the fact that the amendment of the Criminal Act of 1992 contains some new categories of criminal offence in relation to the environment, the majority of violations in the sphere of environment remain in the misdemeanour category. Substantially milder criteria are applied when judging a criminal offence in the sphere of environment than in other spheres.

The Civil Law Procedure Code (No. 99/1963) provides that all persons have the right to vindicate their legal rights in a court of law. Under Article 202, rights of appeal are limited in certain cases. In 1990, a provision was added to the Penal Code making certain acts a crime if they damaged or would potentially damage the environment.

After November 1989, the General Prosecution also mobilized and declared its interest to participate in the matters of environment. It also established a special department. The Ministry of Control was even more engaged in this area, but it disappeared in 1992. It was replaced by the Supreme Control Office which has different powers.

Environmental policy issues

Environmental Legislation

The Slovak Constitution (adopted by the Slovak Parliament in September 1992) includes several progressive statements concerning environmental protection, as well as the rights and duties of different subjects to the environment.

The first environmental law – the Nature Protection Act – was introduced in 1955. Other legal norms adopted during the 1960s were: the Forest Act, the Act on the Health of the People, and the Air Pollution Act. These laws were relatively progressive but the number of exceptions from enforcement which were granted rendered them virtually ineffective, as was the case with the Water Act. Any environmental law, no matter how progressive and strict, operated under extremely unfavourable conditions. During the Communist era, no Ministry of Environment or the like was established. Responsibility for the environment was dispersed among other ministries. Ministries were responsible for the state of those aspects of the environment with which they were somehow economically or administratively associated (Jehlička and Kára, 1994).

A summary of the development of relevant legal structures and legislative processes from 1990 to 1994 follows below.

In the period before the split of Czechoslovakia, several new environmental laws were enacted at both the federal and Slovak Republic level. One of the most important legal norms is the generally applicable Environmental Act (No. 17/1992), which is based on the principles of sustainable development, prevention, prudence, punitive measures, emission and immission limits. Both general rights and duties are defined – it is declared, *inter alia*, that everyone who learns about a threat to the environment or about environmental damage is obliged to take such measures that are within his or her power to eliminate the threat or minimize its consequences and to report the fact without delay to the state administrative authorities.

In 1991, the Waste Management Act was approved – the very first legal norm of this kind in the history of the Czechoslovak legal system. It provides for effective means to lower the production of waste and defines essential duties for waste treatment in compliance with the regulations of Member States of the EU and international conventions. This law represents a set of complex legal measures based on the principle of responsibility of the originator and his or her duties to recycle waste and to implement waste-free technologies. Two other Republic laws dealing with waste management processing aspects were adopted during spring 1992.

The goal of the Federal Clean Air Act (309/1991) is to stop and gradually lower air pollution. Regarding the relationship between the state administration and the public it is stated in this Act, *inter alia*: '. . . the appropriate governmental authorities are required to make accessible full and timely information about air quality and about the specific contributions of individual sources to air pollution . . .'. This obliges operators of large and medium air pollution sources to inform the public in an appropriate way in case of serious and imminent danger or impairment of air quality.

After the split of the Czechoslovak Federation, all mentioned Acts were formally adopted by the new independent Republics. These federal laws provided the basis for laws issued by the Slovak National Council, which further elaborated upon the main principles and mechanisms of the federal laws and defined the specific modifications for the Slovak Republic. Laws to be amended include:

– Regulations specifying emission and immission limits (Law 134/92 on state management of air protection).
– Law 311/1992 on penalties for air pollution.
– Statement of the SCE 407/92 stipulating the list of classes of pollution sources and the list of pollutants and their limits. This Statement stipulates details in setting emission limits for existing sources of air pollution. It also includes regulations in the area of waste management.

In April 1994, more than two years later than the Czech Republic, the Slovak Parliament adopted the Environmental Impact Assessment (EIA) Act. The Act is a system for evaluating the environmental impacts of new construction. It is an important means for preventing pollution. In a general way, this already had been declared in the Environmental Act. Only with the EIA Act was it put into a concrete and executive form. A long time was needed in Slovakia for the preparation of this Act. In comparison with the majority of similar foreign laws, it includes more positive features. For example, paragraph 35 states that all principal development plans, policies and conceptions on both the national and regional levels are to be subordinated to the EIA procedure. There are also some weak points, especially with respect to public participation.

In September 1994, the very outdated Nature Protection Act was updated. The previous Nature Protection Act had existed since 1955. It was a set of basic proclamations joined by a common philosophy. All problems, related to nature protection, which were not covered by this Act, were covered by many different prescriptions and regulations each with different legal powers. The 1994 Act united all these documents 'under one roof'. It also reflected the position of Slovakia regarding international nature protection conventions and agreements. A positive feature of this new law is that it does not only address protected areas or species. The law divides the country into different categories according to a nature protection value and regime. On the other hand, nature protection regimes of the National Parks under this Act seem to be less strict in comparison with its predecessor.

Strategy, Principles and Priorities of the Government's Environmental Policy

The first environmental policy (a federal Czechoslovak document) was called 'State Programme of Protection of the Environment for Czechoslovakia'. It was approved in April 1991. The Programme set guidelines for relevant state institutions, a strategy for legislative work and environmental policy development and made proposals for handling environmental research, education, regional environmental problems and international co-operation. The author of this policy and 'supervisor' of its implementation was the Committee for Environment. Josef Vavroušek, a famous environmental research worker and popular NGO activist, was named its chairman and Environmental Minister of the federal government.

The Slovak Environmental Policy is based on the Constitution of the Slovak Republic which proclaims the right to a favourable environment, to complete current information on the state of the environment and the condition of the environment, and an explanation for the causes and consequences of this state. No one may endanger or damage the environment, natural resources, or historical artefacts beyond the limit specified by the law. The State is required by the Constitution to ensure environmental homeostasis, conservation of natural resources, and effective environmental protection (articles 44 and 45). The Slovak economy is based, according to the Constitution, on the principles of a socially and environmentally-orientated market economy (article 55). The Slovak Republic currently owns only mineral resources, underground water, natural curative resources, and water flows (article 4). The Constitution of the Slovak Republic states that the Republic is the exclusive owner of these resources which means that they belong to the nation and are impossible to privatize. The exercise of property rights may not endanger public health, nature, cultural objects, or the environment beyond the legal limit (article 20).

The orientation and priorities of national environmental policy (approved by Resolution No. 619 of the Slovak Government on 7 September 1993 and by Resolution No. 339 of the National Council of the Slovak Republic on 18 November 1993) are presented below. The National Environmental Policy is focused on the following issues:

- Mitigating the negative impact of components of the polluted and damaged environment on life expectancy and public health.
- Preventing the rise of further undesirable and irreversible changes in ecosystems and other damaging phenomena which destabilize the environment and cause a decline in value and environmental destabilization, lower productivity, or habitability of the land.
- Reducing or preventing the growth of the environmental liability in the privatization process. Determination of liability for the environment liability of the units privatized.

- Increasing polluter participation in improving the state of the environment. Increasing entrepreneurial interest in providing environmental products and services. Reduction of the disparity between environmental needs and resources available for the environment and for effective, low-cost measures.
- Creating conditions for transformation of the economic structure from one with high energy and raw material demands to one characterized by conservative and rational raw material and energy use. This includes a higher product to input ratio, utilization of decontamination procedures and modern, environmentally-safe technologies. It also includes modes of transport which meet environmental standards, proper storage of material which extends their life and re-usability and more accurate evaluation of people's work and abilities.
- Greater reliance on non-traditional energy sources (solar, wind, geothermal). Conservation of natural resources, utilization of biological processes in agriculture. Revitalizing damaged forests, river basins, and devastated areas. Greening of towns and countryside. Optimizing land use.
- Increasing public environmental awareness with emphasis on young people and the business sector. Increasing the level of knowledge concerning the state of the environment in the Slovak Republic. Increasing possibilities and measures taken to improve it.
- Closer international co-operation in the field of development, environmental protection, and sustainable development. Fulfilment of international environmental legislation commitments.

In accordance with these criteria and the environmental situation, the national environmental policy is orientated towards eliminating the causes of air, water, and soil pollution. The quality of these elements of the environment most significantly affects the other components of the environment by determining the state of the environment and directly or indirectly impacting all forms of life. Closely related to this is the influence of the level of technology used for minimizing the effect on the environment and, in particular, minimizing waste production.

Therefore, the following areas have been designated as the priorities of national environmental policy:

- Global environmental security and protection of the atmosphere against pollutants.
- An adequate supply of drinking water and reduction of water pollution to acceptable levels.
- Soil conservation and the purity of foodstuffs and other products.
- Proper disposal or utilization of waste and minimization of its production.

– Preservation of biodiversity, conservation and rational use of natural resources, and optimization of land use.

The above priorities are comparable to those of the states with which we share the longest borders, namely the Czech Republic, Hungary and Poland.

Education

Environmental education belongs to the sectors where positive changes were anticipated by the environmentalists. Despite the awareness about environmental coherencies which increased importantly during the second part of 1980s, the quality of the environment does not reach the necessary priority in the ranking of the human value system, which is also the result of an underestimation of the environmental education. Ecological education was included in the official education programme after November 1989, but it missed conception and co-ordination. It was shallow. Pedagogical staff were unprepared, and this hindered achievement of the expected effects. Environmental education was used for mechanical delivery of knowledge – not for practical training through the use of model situations or real problems. In the most important period of personality formation (the pre-school years), education in responsibility for the environment is almost totally absent.

Official environmental education was substituted for many years in Slovakia by unconventional, informal education presented by various volunteer groups. Today this function is performed mainly by the Slovak Union of Nature and Landscape Protectors, the Tree of Life Movement, and the re-legalized Scouting movement.

In 1991, a cross-sectoral Environmental Education Advisory Body to the Minister of Environment was created. The main aims of this body were promotion of implementation of environmental education and co-ordination activities of different relevant bodies in this field. In 1992, the Nature Sciences Faculty of Comenius University, in co-operation with the above-mentioned NGOs, created the National Centre of Environmental Education. The activity of this Centre was limited from the beginning by lack of finance and professional staff. It lasted for two years.

A pan-Slovakian network of environmental education centres (ecocentres) also was established. Each has its own buildings, equipment and professional or semi-professional staff. The centres, which create a lot of work in the field of environmental education, also have permanent troubles with money, lack of public interest and administrative barriers.

The contemporary government gives a low priority to environmental education. In contrast to the period from 1991 to 1992, there does not exist any department or any full-time professional who deals with environmental education either in the Ministry of Environment or in the Ministry of Education and Science.

Information

In many cases it is much easier to get information from foreign sources. Problems with information arise from its dispersion and non-complexity. There is problematic access. Very often, existing environmental data and information have low reliability. Institutions keep information secret. They are unwilling or do not have the ability to provide information. Despite the right to information, which is a part of the Environmental Act, there are no corresponding regulations at the national level and no possibilities for judicial enforcement of this right. Governmental bodies very often ignore demands of NGOs and the public for information.

The mass media, with small exceptions, is incredibly passive regarding the revelation of trespassing in the area of the environment. They even lag behind the interest and qualifications of the artificially limited 'eco-journalism' that existed before November 1989. Unilaterally offered information, e.g. about the Gabčíkovo Dam issue, reminds one of the state propaganda of the 1950s. Power interests and directly or indirectly compromised TV and radio support the condition of a low civil conscience. There is not only a lack of information, but any information is biased, filtered, deformed and silenced. Information is presented through schematic approaches and use of demagogic 'arguments'. TV and radio are subordinated to the Government and lobbies. Environmental, alternative and independent programmes and information, in general, are limited.

International Relations and Co-operation

For Czechoslovakia and other countries of the 'Eastern Bloc' the typical feature from 1948 until 1989 was isolation from Western countries. This had implications also for the environment. Czechoslovakia was not a signatory to many essential treaties on the protection of nature, environment, and natural resources. The approach of central authorities to international treaties changed after the Revolution with the establishment of national environmental authorities and with the creation of the Federal Committee for Environment in the early 1990s. The Committee co-ordinated numerous negotiations by governmental and non-governmental organizations on bilateral and regional co-operation.

The majority of international conventions dealing with the environment were signed by Czechoslovakia or Slovakia in the last five years. The Convention on Environemtal Impact Assessment in a Transboundary Context (Espoo, 25 February 1991) is not effective yet because bilateral executive agreements between Slovakia and other countries are missing. It is a very promising tool for solving and preventing acute international environmental problems.

The Conventions and Declarations from the Rio Earth Summit are among the most important international Conventions which Slovakia has signed.

However, the implementation of such documents as the Rio Declaration or Agenda 21 is disputable because of missing national legislation which allows implementation. This occurs in several areas:

- Involvement of the public, self-governments, independent experts and NGOs in the decision-making process related to environment and development is even weaker than before the Earth Summit.
- Support of unsustainable activities like nuclear power plants, big dams and highways is even stronger than before the Earth Summit.

The first Pan-European Ministerial Conference 'Environment for Europe' – a meeting of the representatives of states and international organizations held in Dobřiš near Prague from 21 to 23 June 1991 – was a very significant step in the Helsinki process and also a preparatory step to the UN Conference on Environment and Development (UNCED) held in Brazil in 1992. The meeting at Dobřiš initiated closer international co-operation and higher integration of the former isolated East European countries into an environmentally-minded international community. At the second Pan-European Ministerial Conference in Lucerne, in April 1993, Slovakia joined the Environmental Action Programme (EAP) for Central and Eastern Europe.

Slovak NGOs had been involved in EAP issues since the beginning of 1993 when a draft of the EAP was submitted for discussion to the international NGO group which was preparing the Lucerne NGO Pre-Conference. A four page critical statement was sent to the Slovak Ministry of Environment. Representatives of some Slovak NGOs were invited by the Ministry to take part in the EAP draft presentation just before Lucerne. Critical comments to the EAP draft and the formal presentation were repeated. There was no real chance to influence the draft by the NGOs and independent experts and no real discussion on the submitted draft. None of the Slovak NGO critical comments was incorporated. In addition, no Slovak NGO representatives were made part of the Slovak Ministerial Delegation to the Lucerne Conference. This was in contrast to some other Eastern European delegations to the Conference who had NGO participants.

Critical statements relating to the draft EAP were repeated in 1993 and 1994. There was no response to these statements. The February 1993 statement said:

> Under the notion Environmental Action Programme, we are understanding something like particularisation of the strategy (shift) to the way of sustainable development in the spirit of Rio/UNCED conclusions (which have been joined also by the CEE countries). Presented EAP aims more or less only at a defensive approach to the limitation of the sources of environmental degradation and it gives minimal attention to the search for a new, effective, healthy society from the point of view not only of the environmental aspects but also the economic and social ones.

We also criticized the static approach of the EAP draft in the physico-geographical or landscape ecological level, as well as in the socio-demographical one. 'The CEE region needs a more territorially differentiated, more geographical, dynamic and complex approach. In the next EAP process, it would be desirable to gather more experts, not only from the economical, legal, technical and medical spheres, but also, and especially, from the biological and social ones, and also from disciplines like geography, landscape ecology, etc.'

We concluded our statement by these recommendations:

- To initiate the creation of the advisory bodies to the Governments, Parliament and Ministry of Environment for particularisation, development and implementation of the EAP (with participation of independent experts, NGO activists, etc., and with increasing stress upon natural and social sciences).
- To initiate the formation of national cross-sectorial committees for the preparation of Pan-European Ministerial Conferences and similar events. Besides representatives of the legislative and executive power, to invite for its members also prominent expert-environmentalists and representatives of NGOs.
- To organise experts' seminars, public hearings, etc., with the authors of the EAP, as well as with persons who are responsible for the next development and implementation of the EAP. To be in permanent dialogue with them.
- To elaborate a shorter version of the EAP for the wider public and to translate it in the national languages.

Thanks to the Regional Environmental Centre (REC), the last demand was realized in 1994, but no other recommendation has been accepted by the Slovak Government yet. There has been no answer to our proposals and demands.

In September 1993 the Slovak Government adopted the *Strategy, Principles and Priorities of the State Governmental Environmental Policy*. It was approved by Parliament in November 1993. The Slovak Ministry of the Environment understands 'Strategy' as a fundamental part of the NEAP prepared in the frame of the Sofia process.

The problem with using the term 'Strategy' to stand for the Slovak NEAP is that the material is two to three years old. That means that the short-term goals in the document are obsolete. In addition, a serious discussion on the EAP, which was adopted by the ministers in Lucerne, as well as Agenda 21 or other relevant international documents, is missing in the Policy. If these documents are not mentioned in the national Policy, it means technically that they do not exist. The 'Strategy' could stipulate how the National Environmental Action Programme would be developed. The idea behind the recommendation from the World Bank was that nations would not adopt the Environmental Action Programme as their policy but instead would construct a policy which combined parts of the EAP which were applicable with their own environmental experience. The 'Strategy' ignores not only the NEAP but the other important international documents as well.

Another important idea in the EAP/NEAP process is that there be active co-operation between governmental/ministerial bodies, the independent experts, NGOs and the general public. This has not existed in the process of the Strategy preparation and in the NEAP process. The official MOE statement from March 1995 says, 'The mechanisms for public participation in NEAP preparation have not been adequately developed. . . . Putting public discussion at the end of the process is inadequate . . .'. Another issue of concern is that the distribution and use of about 1 million ECUs from the European Union for supporting NEAP preparation in Slovakia is not transparent.

In February and March 1995, we sent letters together with brief questionnaires dealing with the NEAP issues in Slovakia to 22 Slovak NGOs as well as to the Minister of Environment and the Parliamentarian Committee for the Environment and Nature Protection. We received three written answers from NGOs and none from the officials. The NEAP should have been submitted for governmental approval by September 1995. There seems to be no interest in its development either from the majority of NGOs or the Government. The Ministry of the Environment would like to give the illusion that there is interest in co-operation. The NGOs do not believe there is real opportunity for meaningful participation in the process.

7.3 PUBLIC PERCEPTION AND PARTICIPATION

Public opinion on environmental issues

Just after the 'Velvet Revolution', there were two main priorities generally accepted by the Czechoslovak public: human rights and restoration of freedom, as well as improvements in the state of the environment. Trends signalling that environmental and/or sustainability dimensions could belong to the main dimensions of the post-revolutionary development in Slovakia matured during the 1980s. The 'Velvet Revolution' brought all these together.

First of all, there were catastrophic trends of development in practically all spheres of the environment, as well as in the spheres of health, morbidity and mortality of the population. Recognition of these problems started to penetrate more intensively from the professional and political circles to the public in the second half of the 1980s. Conceptual materials such as the *Cumulative Economic and Social Prognosis of the Czechoslovak Republic and the Slovak Republic* and other prognostic studies from a number of teams were released. There were publications from informal experts' groups, e.g. *Bratislava/aloud* (1987), which were issued by the Town Organization of the Slovak Union of Nature and Landscape Protectors (SZOPK) in Bratislava. *The State and Development of the Environment in Czechoslovakia* (1989) issued by the Ecological Section of the Biological Society of the Czechoslovak

Academy of Sciences had an effect. These publications described in a very open and complex way state and development trends in the sphere of the environment. They discussed the responsibility of the whole society and the crisis of the system which caused such conditions.

In Slovakia, and above all in Bratislava, these groups succeeded in creating a strong and potent environmental movement able to generate an extraordinary broad scope of activities on a philosophically consistent basis. This movement attracted a number of outstanding personalities who were ready to overcome typical Slovak tendencies of individualism or non-co-operativeness. These personalities wanted to collaborate without competition over the interests of their own organization. They were interested in the whole society. The attributes of this movement were altruism, charisma and spontaneity.

Organization was through the existence of an environmental NGO with a long tradition of legality and relative independence, the SZOPK – Town Committee in Bratislava. It attracted into the community of environmentalists people who under normal democratic conditions would be successful in their chosen professions or in other non-environmental movements or civil initiatives. Besides regular members or activists, the environmental movement of the 1980s had a broad coalition of sympathizers and supporters at its disposal, especially in the intellectual community. This fact became evident during the controversy over issuing an environmental pamphlet with political implications – *Bratislava/aloud* – when public support proved to be overwhelming.

Negative perspectives of development trends in the Slovak economy, social sphere and culture signalled most expressively a need for change with positive environmental implications. Change was perceived as being needed to allow the following:

- Decentralization
- Conversion of the armament industry
- Reduction of the metallurgy industry (mainly coloured metallurgy), heavy chemistry and machinery
- Strengthening of finalization (final productions)
- Rapid increase of energy efficiency of the Slovak economy
- General reduction of the transport costs per unit of GNP
- Radical reduction of the extent of buildings under construction
- Substantial increase of the production quality including environmental properties of products, also in connection with their market competitive ability
- Improving utility values and aesthetical architectural–urbanistic qualities of the new urban and rural buildings and constructions
- Making the care of historical monuments more effective.

These and other needs for change corresponded with the need for improving the quality of the environment.

Environmental problems and the environmental movement had a freedom as compared to more tabooed and more proscribed themes. Maybe it was also the influence of the Soviet 'Perestrojka' which at its very beginning sanctioned giving vent to social problems even through the environment. That is why in the time of the 'Velvet Revolution' and shortly after it, there was a strong feeling that change was needed even in the environmental sphere. The new Green Party was highly preferred and had good support from the population according to opinion polls.

There was a concentration of civic engaged personalities in and around the environmental movement. They enjoyed general public support in the groups of better informed citizens. SZOPK had a relatively good infrastructure. It had contacts with the mass media, and printing equipment was available. For these reasons, the revolutionary movement of VPN (Public Against Violence) was able to organize. The Revolutionary Secretariat, for instance, in the first weeks after 17 November was identical to the SZOPK Bratislava Town Committee. More than half of the first protagonists of the 'Velvet Revolution' in Slovakia, as well as of the participants in the first independent TV and radio dialogues, were members of the environmental movement.

As a result, members of this movement became members of the reconstructed governments and parliaments. Enormous interest was expressed by foreign guests, mainly journalists, environmentalists and greens. There was a sudden opening of space in the mass media which brought publicity, popularity and social respect to the environmental movement. There was a certain equivalency attributed to environmentalism and revolution.

Until the end of May 1990, the bad state of the environment was the most frequently mentioned problem according to the opinion polls. The solving of this problem was considered top priority. In October 1990, the growth of nationalistic and socio-economic problems replaced the environment in its number one position. From this time on, environmental issues have continuously decreased in the polls. In May 1990, 20% of the inhabitants of Slovakia considered the environment the most important problem. In October of the same year the number had fallen to 7%.

The reasons for gradual retreat of the environment and/or the environmentalists from their positions were paradoxically partly a consequence of the reasons of their pre-revolutionary and revolutionary significance and influence. The actual freedom of assembly and speech as one of the first real results of the Revolution allowed a number of activities and created a broad environmentalists' movement. Different activists were able to create their own, independent, specialized associations of interest.

During the political discussions about dividing power, the Green Party was more and more discriminated against. In spite of that, it remained between second and fourth choice of election preferences according to public opinion surveys until six weeks before the parliamentary elections. During the election campaign, the Party lost because of its insolvency and because of discrimination and demagogical accusations on the part of its political rivals. There was a relative lack of outstanding personalities. There was also a certain amount of amateurism inherited partly from the environmental movement. The seats in the first really democratic parliamentary elections in Slovakia were divided among the VPN, Christian Democrats, Communists and Nationalists. The Greens received only six seats out of a total of 150 in the Slovak Parliament and none in the Federal. A similar result was repeated in the municipal elections in autumn 1990.

As a result of these events, the environmentalist movement began to suffer demotivation and deactivation. SZOPK still remained as a politically independent, non-governmental and non-profit organization. In a time of an information (often rather mis-information) explosion in Slovakia, information provided by environmentalists was overshadowed by other information and problems, often of a tabloid character. In Slovakia, in general, political and economic worries and fear of mass unemployment inevitably came to the foreground. Unemployment had been unknown in the past decades as the right to work was guaranteed by law. This was another reason for the slackening of interest in environmental issues and for the growing indulgence for the polluting employers. The notorious non-solving of environmental problems (including the most affected regions) resulted in a growing discontent of the local residents, though such discontent did not lead to action.

Public participation in environmental decision-making

In the Constitution of the Slovak Republic, the rights and obligations indicated in Box 7.1 are provided with respect to public participation.

In Slovakia, voters have no right to initiate a legislative process. However, voters do have a right of referendum. A minimum of 350 000 voters may petition to call a referendum on any law (except for laws pertaining to basic rights and freedoms, taxes, inland revenues or the state budget). Local referendums must be held on a petition of a least 20% of voters of a municipality. The Government sometimes seeks public opinion on draft laws (this happened for instance in 1993, when independent NGO experts contributed to the process of drafting a new Nature and Landscape Protection Act).

Provisions for right to information in relation to the environment as covered by other laws are given in Box 7.2. The general problem with respect to these provisions is that procedures and measures for their implementation are largely lacking in practice.

The Clean Air Act obliges operators of large and medium air pollution sources to inform the public in an appropriate way in the case of serious and imminent danger or impairment of air quality. The reporting is required also by the waste management programmes and some reporting requirements are included also in the Water Act. However, reporting is required only to the water management authorities and not to the public. Articles 17 to 19 of the

---

Box 7.1 Rights and obligations in the Constitution of the Slovak Republic (adopted in September 1992) with respect to public participation

- *Right to Healthy Environment.* 'Every person has a right to complete and current information on the condition of the environment and the causes and consequences of this state.' (article 45)
- *Right of Expression.* 'The freedom of expression and the right to information are guaranteed.' (article 26 (1))
- *Right to Information.* 'The State central authorities and the authorities of regional administrations must provide information about their activities in a reasonable manner and in State language. The conditions for this paragraph will be set by law.' (article 26 (5))
- *Rights of Free Assembly and Free Association.* (articles 28, 29 and 37)
- *Right of Petition.* 'The right to petition is guaranteed. Every person, either individually or in collaboration with others, has the right to petition the state authorities and the authorities of regional administrations, to submit proposals and to lodge complaints.' (article 27 (1))
- *Government's Relationship to the Citizens.* 'The citizens have the right to participate in the administration of public affairs either directly or through the free election of their representatives.' (article 30 (1))
- *Right to Petition to the Constitutional Court.* 'Citizens may petition the Constitutional Court for review of legal decisions (action of official) which violate the constitutional rights and freedoms of citizens.' (article 130 (1f))

General Environmental Act oblige citizens (article 19): 'Everyone who learns about a threat to the environment or about environmental damage is obliged to take such measures that are within his or her powers to eliminate the threat or minimize its consequences and to report the facts without delay to the State administrative authorities.'

Several additional provisions with respect to public participation exist in other laws. Some of the most important are:

- The *Law Establishing the State Administration for the Environment* (No. 595/1990) requires environmental offices to co-operate with other bodies of state administration and municipalities. They can also co-operate with non-governmental organizations and citizens' initiatives for environmental protection.
- In the most explicit and detailed way, cases and forms of public participation are elaborated in the *Act on Environmental Impact Assessment* from 1 September 1994. The EIA Act makes the position of the public in the assessment process equal to the position of other subjects. The public is represented by: (i) different 'physical' and/or 'legal persons' or their groups; (ii) concerned communities; (iii) civil initiatives; (iv) civil associations.

    Division II of the EIA Act deals with civic initiatives and the participation of civic associations. According to paragraph 9/1 'For the purposes of the Act a "civil initiative" means not less than 500 physical persons

---

Box 7.2   Provisions for right to information in relation to the environment by various Slovakian laws

The *General Law of the Environment* No. 17/1992 provides that 'Everyone has the right to true and accurate information about the state and development of the environment, the causes and consequences of the state, activities which are being prepared and which could change the environment, as well as to information about measures taken by the authorities responsible for environmental protection in order to prevent or remedy environmental damage. A special regulation may stipulate cases in which such information can be restricted or withheld.' (article 14)

'Everyone may approach the relevant authority and claim, in a prescribed manner, his or her legal rights stipulated by this Act and other laws and regulation concerning the environment.' (article 15)

The *Administrative Law Procedure Act* (No. 71/1967) includes a provision allowing the local governments to give information to persons who show 'deep interest' in a matter which is being decided upon.

By the *Law on Organizational Structure of Ministries* (No. 453/1992) the state administration is responsible for making environmental information available to the general public. Similar formulations are found in the *Law on State Administration of Waste Management* (No. 494/1991).

By the *Clean Air Act* (No. 309/1991) and the *State Administration of Air Protection Law* (No. 134/1992), the appropriate governmental authorities are required to make accessible full and timely information about air quality and about the specific contributions of individual sources to air pollution. This law also requires county environmental offices to make information accessible to the general public.

---

older than 18 years, including at least 250 persons with a permanent abode in the municipality affected, who sign a joint statement on a proposed activity.' According to paragraph 9/4, 'If a group of not less than 250 physical persons of more than 18 years of age, including at least 150 persons with a permanent abode in the municipality affected, sets up a civic association, e.g. environmental NGO (under special provisions, for the purpose of protecting the environment from the particular activity assessed under the Act), then this civic association shall also take part in the administrative hearing in which a decision is reached on the permission of the given activity under the special regulations.'

There still is a lack of practical experience with public participation in the EIA process. EIAs are connected mainly with highway and water dam projects. The greatest public participation was provoked by the EIA of the Mochovce NPP finalization, organized by EBRD in 1994 and 1995.

According to paragraph 35 of the EIA Act, a proposal for a substantial development policy and territorial planning documentation must contain an assessment from the point of view of its presumed impacts on the environment and, if necessary, also a proposal for measures to eliminate or reduce the adverse impacts. Public participants (NGOs) have appeared in connection with comparing the official Water Management Policy with alternative ideas from NGOs, as well as in connection with an EIA on the proposed Actualization of the Energy Policy.

– Under the *Act on Physical Planning and Building Code (Building Act)*, parties to the land planning decision and investment permitting processes have the right to require inspection of facilities prior to their completion and subsequent to final approval of completion.

Under Article 1, paragraph 2, local land-use plans are required to be developed with specific public consultation procedures. Notice of the public participation process must be given according to the usual means of notification in the locality, such as through the mass media. The draft plan itself must be available for public inspection for 30 days. The public has an opportunity to comment upon the plan. The comments of those whose rights to real property are not affected by the plan are taken as an indication of public opinion; no specific response is required to these comments. Persons whose rights to real property are affected by the plan have the right to receive written answers to their objections from the planning authority, which is obliged to indicate how the authority has made adjustments to the plan to reconcile the objections or the reasons why the objections did not cause a change in the plan.

Under the Building Act, persons whose property rights may be directly affected by a land-use planning decision have the right to participate in such decision-making through the EIA procedure. In such a process, municipalities are also interested parties. They may act in their own capacity and as representatives of the broader public when the proposed decision relates to large construction works, such as a dam or nuclear plant.

In spite of the above-mentioned number of Acts and procedures dealing with public participation issues, the real situation in public participation is far from optimal. Decades of passivity of the population, interrupted only by short periods of enthusiasm and public activation (1968, 1989–92), have created an atmosphere of ignorance of both civic rights and responsibilities. This feeling has been joined strongly with the tradition of state 'paternalism' in which transmission of all duties and responsibilities flows from the State (government) to the individual.

In this situation, NGOs play a much more important role than public and/or civic initiatives. In early 1990, Slovak environmental NGOs organized a living chain between Hainburg in Austria and Gabčíkovo in Slovakia with about 60 000 participants. Many other direct actions have been organized between 1986 and 1992 against completion of the Water Works Gabčíkovo–Nagymaros. Greenpeace, Children of the Earth and other NGOs organized dozens of direct actions and petitions against nuclear power plants in Slovakia. The Slovak Union of Nature and Landscape Protectors (SZOPK) organized several 'silent demonstrations' against the Poprad–Tatry candidature for the

Winter Olympic Games 2002 in the Tatry National Parks. Some environmental NGOs are active in organizing demonstrations against the use of chlorofluorocarbons and freon gases. The NGO 'Freedom of Animals' in 1994 was successful in opposing cruelty to animals through basic school education and training. Probably the greatest common action of the Slovak NGOs (not only environmental ones) was a common statement by 83 of them against an EBRD loan for completion of the Nuclear Power Plant in Mochovce in early 1995. NGO activists also have been active in the process of public participation related to the Mochovce case.

Environmental issues with a political background played a vital role in public mobilization at the end of the 1980s. In spite of Communist censorship before 1989, a Club of the Ecological Journalists in Bratislava was created with some 30 members. The environmentalists used to organize common actions with the journalists (e.g. excursions to endangered localities). Many members of the Club also were members of environmental NGOs.

This situation not only reflects the activities of the environmentalists but also the fact that real political discussion and an independent sector did exist in that period and that environmental issues and discussion about them were not under such strong control in comparison to many other critical issues. After the 'Velvet Revolution', environmental issues started to be over-shadowed by other new and unexpectedly opened issues. However, the reason for the drop of mass media interest in the environment is not fully under-standable. There is no general lack of interest about such critical, interesting and attractive issues as nuclear power plants, large dams and/or other large constructions, alternative energy or tourism, sectorial or communal develop-ment plans, huge international events like the Rio Summit or results of sociological surveys of ministers and MPs' opinions on different environ-mental and/or environmentally-related issues.

7.4 MAJOR PROBLEMS AND OBSTACLES

Major problems confronting environmental protection

Problems and obstacles which block improvement of the environmental situation occur practically in all spheres of the life of society.

In the administrative sphere, transparency disappeared especially on the lower administrative levels where originally united national committees were divided in 1990 to several species of state, municipal and local administration. (Before 1990 district and local self-governments and district and local state administrations of all kinds had been joined under one title, the District (eventually the Town or Local) National Committee.)

In spite of the origin of the Ministry of Environment as a central organ of state administration, the division of powers between the various departments involved in environmental decision-making is not always logical and readable.

The inner structure and division of powers in the Ministry of Environment can be similarly characterized. The responsibilities of the Ministry of Environment are incomplete, and there are even efforts to cut down its present responsibilities. The situation is further destabilized by frequent changes of staff and changes to the organizational structure brought by practically every new government (from 1990 to 1992 there were four governments; in 1994 there were three). When political instability is connected with personal instability, 'political qualifications' often become more important than professional ones. Such problems occur on the national but also on the local (regional) levels.

Some of the *financial issues* were discussed in Section 7.1. Here, the main problems are summarized briefly. There is an underestimation of the environmental sector within the framework of the distribution of means from the state budget. If one calculates the allocation for environment of the state budget per employee or per processed item of the agenda of the Ministry of Environment, then the budgetary allocation for environmental issues is the least financially supported. Every year, the state funds, especially the State Environmental Fund (SEF), receive fewer subsidies from the state budget. On the other hand, there is no real success in rebuilding the SEF into a so-called 'revolving fund'.

The 'polluter pays' principle is not sufficiently applied to state property. State owned companies, often with very polluting production, are in many cases insolvent, and eventually in primary and secondary indebtedness. The modernization, conversion, and 'ecologization' of production proceed slowly because of the pace of transformation (mainly due to the slow privatization, a purposeful delay of the effect of the Act on Bankruptcy and highly subsidized prices of energy by the State). Legislative gaps in the sphere of economic tools – incentives and their inconsequent application, numerous other factors of a non-economic nature, as for instance the insufficient authority of control organs, and a decreasing belief in the possibility of change – have negative effects.

Foreign assistance is insufficient and there is a considerable gap between optimum expectations and results. As a rule, foreign assistance is orientated more to unsustainable than to sustainable projects. Collaboration on acquisition, redistribution and implementation of financial resources for the state and public sector is not sufficiently transparent (in connection with the general inefficiency of the state sector and lack of real public control of it). The use of foreign assistance in the state sector, when compared with assistance in non-governmental and non-profit organizations, is not efficient enough.

In spite of numerous *legal norms* adopted in the sphere of environmental legislation, the legal system still has many drawbacks. The Ministry of Environment does not consistently cover the environmental problems in all relevant spheres. Some obsolete and outdated Acts like the Water Act or the Act on Territorial Planning and the Building Order have not been amended yet. Other laws are being gradually 'softened'. There is a retardation of the effective date for environmental regulation especially if pollution limits are concerned.

In general, practical experience with environmental legislation, its implementation and the enforcement of environmental policy are missing at all levels of state administration (management). Also missing is a clear code of conduct for foreign companies and support programmes for small companies willing to invest in the environmental industry and its markets. Slovakia obviously misses environmentally-orientated consultancy and advisory centres for both the public, self-governments and/or businesses. Especially missing are similar centres for promotion of sustainable technologies and ways of life, as well as development programmes.

Environmental education at schools was implemented only in a limited and often formal way. Curricula, qualified teachers and an idea about what environmental education really is are missing. A significant contribution in the area of environmental education still is represented by non-governmental organizations. But the possibilities here are also very limited because of lack of finances, personnel, interest and artificial barriers. Official cultural facilities that also played some role in this area before 1989 mostly have disappeared due to the lack of means and interest.

The *media* deeply underestimate environmental issues. The circle of journalists who consider themselves environmental is even smaller than before 1989. The key media, especially Slovak Television, are dominated by unilateral support for technocratic intentions.

An overall *shortage of relevant information* represents an absolute and relative problem. It is absolute in the sense that a sufficient amount of information and knowledge about complex environmental questions still does not exist. Some areas of interest, for instance in the field of health care statistics, are not observed at all. Though the amount of information grew under the effect of the opening of the borders and freedom of speech, the availability of information is hindered by other barriers. Rather than ideological censorship, there is now commercial inaccessibility for those who depend on information on a non-commercial base, e.g. persons working in research. The relative shortage of information is caused also by the fact that a great part of it is interpreted tendenciously. Too much information is disinformation or information noise.

From a NGO perspective, the following philosophical and spiritual barriers to implementation of an effective environmental policy were identified by Zamkovský (1992):

1. Political barriers
    (a) Lack of synchronization in the formulation of general policy.
    (b) Weak personal potential and structure of the Ministry of Environment.

(c)   Incomplete legislation.
(d)   Lack of pro-environmental economic instruments.
(e)   'Cabinet' (non-transparent) decision-making.
(f)   Insufficient integration of the Slovak environmental community.
(g)   Missing 'green' media.
(h)   Missing environmental education and awareness.
(i)   Lack of grassroots organizations and lack of management and citizens' social change skills.
(j)   Unstable political and complicated social situation.
2.   Technical and technological barriers
(a)   Lack of finances.
(b)   Lack of environmentally-sound technologies.
(c)   Lack of adequate know-how.

The way ahead

The present situation is characterized, in comparison with 1989, by progress in the sphere of environmental legislation. During the past five years there has been partly newly adopted and partly amended legislation. However, the process of creation of new environmental legislation cannot be considered finished, and the products of the legislation process of the past five years are not perfect either. In practice, the development after 1989 brought about indisputable positive features but also numerous negative ones. For instance, the drop of electric power consumption in industry was lower than the reduction of industrial production. This demonstrates that an extraordinarily low energetic efficiency, typical of the past decades, even decreased. The yearly rate of state investments into the environment was in the past two years even lower than in the 1985–89 period.

The trend of 'unsustainable' investments known from the preceding regime continued, i.e. investments that were not possible to be realized at that time (expansion of the aluminium plant in Žiar, Nuclear Power Plant Mochovce, development of the car industry, etc.). The import of numerous products, technologies and customs from the West (the boom in automobiles, non-recyclable packaging, tobacco products, efforts to export wastes and obsolete technologies to Slovakia, 'pulp-culture', support of consumerism and an extension of that to life philosophy and practice) had a more negative than a positive effect.

Even if we talk about an improvement of some indices (SO_2, CO_2 scores and some indices of water pollution), we find that it is the result of a drop in production, increased prices of inputs, short-term increase of authority of the environmental offices, partial modernization, gradual introduction of measurement and regulation techniques, experiments with classification and recycling of wastes, control of the technical condition of vehicles, and other measures.

Possible Scenarios – Negative

With the unstable political situation and the unfinished process of economic transformation, it is difficult to predict what tendencies will prevail in the immediate future. Perhaps the *least probable* change will be the comeback of the vehement approach which characterized the 1990–92 period when a decisive number of changes were realized. If the political, economic and social situation does not change, a prolongation of the trend from the period June 1992 until March 1994 can be expected. This period can be denoted as a stagnating one, as far as the implementation of environmental rights and distinct investments into the improvement of the environment are concerned.

An overall marked regress can be expected if the following occur:

– The situation of state enterprises and their legal and transparent privatization stagnate.
– Discrimination against towns and communes through redistribution of state subsidies and fees from polluters in the State Fund of the Environment if they are thought to be disloyal to the central power.
– Credits and governmental guarantees are further absorbed by the development of 'socialist' large-scale production (new dams, power plants, metallurgical plants, etc.).
– Unilateral, anti-ecological propaganda by the key media continues.
– Devaluation of newly adopted legislation.
– Time horizons for pollution limits compatible with the limits effective in EU are readjusted.
– Evasion of effective laws through the introduction of exceptions.

Such trends can be associated with efforts to preserve the life of large polluters without any substantial conversion or modernization. Blockage of new, less demanding, environmentally-friendly, diversified and sophisticated production also occurs. Obviously towns and communes will not have means for the liquidation or the processing of communal wastes, overall hygiene, maintenance and development of urban greenery, improvement of transport systems, etc. Addressing the political, economic and social problems (for instance, unemployment) can result in increased doubts of authority, or even existential justification of the subjects dealing with environment. According to many of the public, who are supported by some politicians and media, environmental protection is responsible for the rise of unemployment and economic problems in general. For many people, 'freedom' to exploit nature is one of the most relevant freedoms.

If this continues, the overall environmental consciousness of the public will further degenerate. The so-called environmental debt will increase, and notoriously unsolved environmental problems will persevere and grow. Life expectancy, already one of the lowest in Europe, will stagnate or decrease.

A *worse alternative* could be caused by the restoration of the situation that existed before 1989, but with considerably higher inner and foreign debt. The economy would not be capable of competition and several times more expensive raw material imports would be required. The consequences of such a development would be a heavy impact on natural resources, environment, health and the population's quality of life.

An even more negative alternative would be if Slovakia's trends of future development become similar to those of the Balkan or Latin America countries, in combination with growing intolerance and inner-social antagonisms, as well as conflicts in relation to neighbouring countries. One of the first consequences of such development would be even greater disinformation on the state of environment, a ban on basic freedoms and 'excommunication' of independent non-governmental subjects including the environmental movement.

Possible Scenarios – Positive

There are alternatives that could be denoted as optimistic and/or optimal. The *optimistic* version is based on the combination of a rational use of home potential, a systematic reduction of the inherited environmental debt, the building of environmental consciousness and a targeted and efficient foreign assistance. In this alternative:

- The intentions of the state environmental policy would be implemented.
- Environmental legislation compatible with that of the EU would be accomplished and its observation would be ensured.
- In economic transformation, environmental aspects would be taken adequately into account, measures would include decisions over the responsibility for the solution of the problems of environmental debt, the introduction of environmental taxes, etc.
- Through education, as well as the media, the environmental aspect would receive such a position that the public's consciousness in this area would improve.
- International collaboration, the flow of financial and other assistance would be directed to priorities reasonably judged and carefully chosen programmes and projects with perspective, applying mechanisms such as the EIA, Least Cost Planning, 'win–win' strategies or Integrated Resource Planning. Such techniques imply the extension of the field of activity and of the resource acquisition of an energy utility to options on both the demand side and the supply side.
- Citizens who regard the environmental agenda as a priority would be present in the Parliament and eventually in the Government once again.

An *optimum* alternative suggests the implementation of the principles and measures quoted with the preceding 'optimistic version', but set into the context of the conception of sustainable living (development) in the sense of the UNCED (especially Agenda 21) recommendations. It also suggests the elaboration and implementation of detailed national and sub-national Environmental Action Programmes based on national, regional and local visions of sustainable living (development). In practice it would mean, for instance, the gradual conversion of the Slovak economy to a more decentralized and sophisticated one, with considerably less demand of energy and raw material, corresponding to the natural, historical and socio-economic conditions of Slovakia. It should comprise a combination of a renaissance of traditional production (and consumption) in combination with adequate high-tech production. There should be the creation of new jobs, the renaissance of traditional small-scale production and crafts, the development of sustainable tourism, and an orientation to production where Slovakia possesses comparative advantages. A revival of the settlements and the general health of the country and its population is also part of this alternative. This alternative requires practical application of the basic principles of sustainability:

- Ecological, including respect for life in all its forms
- Subsidiarity
- Non-violence
- Optimum and considerate use of domestic resources
- Maximum possible efficiency and closed production–consumer cycles
- Gradual, reversible steps at application of development.

Such an approach would also mean:

- Reduction of demands on energy and raw material of the Slovak industry by at least 50%
- Similar reduction in water loss in waterworks systems
- Radical reduction of the dependence on imported raw material, including primary energetic sources
- Gradual reduction of the environmental load
- An improvement in the capacity to produce more competitive products
- More careful management of soil, forests and further life-giving resources
- Growth of the attractiveness of the Slovak landscape
- An improvement of the population's morbidity and mortality indices
- New jobs increased locally
- Renaissance of local culture
- An overall reduction of developmental risks.

This task is ever more demanding as it is in contradiction to the basic attributes of the past 50 years of development. On the other hand, Slovakia has also some prerequisites for such development. The wide diversity of its physical-geographical conditions requires a diversified and decentralized approach. It is limited by its economic potential because of the lack of primary energetic resources and raw material and the disintegration of the COMECON market. The Czech market, within which Slovakia played a specific role as supplier of strategic raw material, semi-finished products and products especially in the sphere of armaments, metallurgy and heavy chemistry, also influences Slovak development.

Because Slovakia has exhausted proper resources of extensive development and inputs from outside, development represents problems (unreliability of the supplies of raw material and energy, rise of prices and cost, export embargoes, etc.). In spite of the relatively low foreign indebtedness, especially if compared with its neighbour countries Poland and Hungary, Slovakia has a very high internal debt in the sphere of the environment, an obsoleteness of funds, and problems in human health.

The limits of its carrying capacity have been reached and passed because of pollution of rivers, health condition of the forests, physical and chemical degradation of soils, and transport load of settlements. Similarly, the limits in the sphere of the centralization of decision powers associated with state-monopolistic relation to production means, with the absence of feedback as represented by de-monopolization, implies that diversity and the democratic character of society have been suppressed.

As a consequence, individual local areas or whole regions are overloaded by economic activities (unregulated industrialization, collectivization and urbanization). There are strong depopulation tendencies and an abandonment of extensive rural regions and historically valuable small towns. The solution must be found in the search for a restoration of lost equilibrium. Similarly, optimum relations must be searched for between:

- Various dimensions of development (economic, environmental, social . . .)
- Natural and socio-economic systems
- Centres and their hinterland
- Productivity, homeostasis and sustainability of landscape systems in general.

A minimum condition for reaching the above-mentioned intentions will be:

- Respect for the basic principles of sustainability
- Political stability which ensures democratic principles
- Observation of effective laws including compulsory international conventions
- Production respecting the limits of carrying capacity

- Internalization of impacts on environment into the prices of products
- Continuous increase of efficiency in production
- Consumption based on the principle of reasonable sufficiency
- Imposing embargoes on unsustainable products and technologies and patterns of behaviour
- Guaranteed universal and timely information with public participation in the prevention and solution of environmental problems.

REFERENCES

Belčáková, I., Gindlová, D. and Petríková, D. (1994) Slovakia. In: *Manual on Public Participation in Environmental Decision-Making*. REC, Budapest.

ČSOP (1989) *Stav a vývoj životního prostředí v Československy (The state and development of the environment in Czechoslovakia)*. Ekologická sekce Československé biologické společnosti pri ČSAV and Ekologická pracovní skupina Rady ekonomického výzkumu ČSSR in cooperation with ČSOP in České Budějovice and Praha.

Dianiška, I. et al. (1990) Spoločenské problémy očami obyvatel ov (Social Problems reflected by inhabitants). *Ochranca prírody XIV MV SZOPK*, Bratislava.

Huba, M. (1992a) Slovakia's brief environmental characteristics. In: J. Gregor, M. Huba and J. Zamkovský (eds), *Message for Rio*. SZOPK, Bratislava.

Huba, M. (1992b) On a chance to create an ecologically sustainable society. In: J. Gregor, M. Huba and J. Zamkovský (eds), *Message for Rio*. SZOPK, Bratislava.

IUCN, UNEP, WWF (1991) *Caring for the Earth. A Strategy for Sustainable Living*. Gland.

Jehlička, P. and Kára, J. (1994) Ups and downs of Czech environmental awareness and policy: identifying trends and influences. In: S. Baker, K. Milton and S. Yearley (eds), *Regional Politics and Policy – An International Journal*, Vol. 4, No. 1, Frank Cass, London.

Macek, K.J. and Urbánek, V. (1994) *The Environment Industry and Markets in the Slovak Republic*. BMT Environment Ltd., Paris.

Ministry of Environment of the Slovak Republic (1993) *Strategy, Principles and Priorities of the Governmental Environmental Policy*. Bratislava.

Moldan, B. et al. (1992) *National Report of the CSFR (for UNCED)*. CSAS, FCE, Prague.

Plamínek, J. (1993) Postoje veřejnosti k Životnímu prostředí (attitudes of the public towards the environment). In: P. Nováček and J. Vavroušek (eds), *Lidské hodnoty a trvale udržitelný způsob života (Human attitudes and sustainable living)*. STUŽ Katedra ekologie PFUP Olomouc, Olomouc.

Regional Environmental Centre (1994) *Strategic Environmental Issues in Central and Eastern Europe*, Vols 1 and 2. REC, Budapest.

Statistical Office of the Slovak Republic (1994, 1995) *Statistical Yearbook of the Slovak Republic 1993–1994*. Bratislava.

SZOPK (1990) *Slovenské lesy – otázka existenčná (Slovak-forests – a question of existentional character)*. Bratislava.

Šíbl, J. (1993) *Damming the Danube*. SZOPK, SRN, Bratislava.

Zamkovský, J. et al. (1992) Environmental policy in the Slovak Republic. In: J. Gregor, M. Huba and J. Zamkovský (eds), *Message for Rio*. SZOPK, Bratislava.

CHAPTER 8

Conclusions

Patrick Francis, Jürg Klarer and Bedřich Moldan

The previous chapters of this book have presented a wealth of information and insight on the environmental challenges faced by countries of Central and Eastern Europe. Though some problems are shared throughout the region, each country is faced with its own unique set of specific issues to address and each is forging its own path through the transformation period. In this final chapter the authors offer some observations and commentary based on the information presented in the previous chapters as well as their own knowledge and experience in the region. In no way is this chapter intended to be a comprehensive summary of the previous text, nor is it an attempt to identify some ultimate truth about the environmental challenges of Central and Eastern European countries. What the chapter is intended to do is to highlight certain issues and illuminate certain perspectives found to be of particular importance and interest to the authors. For a truly comprehensive review readers are urged to read the book as a whole so that they may draw their own conclusions in an informed fashion.

As the specific circumstances of the different countries vary, so too do the information and viewpoints presented in the individual chapters which must, to some extent, also be seen as a reflection of the varying knowledge, experience and perspectives of the individual authors. A principal objective of the book was to present as objective and accurate information as possible in tandem with the unique insight and commentary of respected experts who are deeply involved in environment and development issues in Central and Eastern Europe. The authors, mostly Central and Eastern Europeans, represent a broad diversity of competencies and interests, and include former Ministers of Environment, leading environmental activists and policy and economics specialists. The different competencies and interests of the authors can, sometimes readily, be detected in the different texts and help to account for the different styles of the chapters. The style of this chapter is intended to be more free-flowing and perhaps provocative, as the authors offer their

The Environmental Challenge for Central European Economies in Transition.
Edited by J. Klarer and B. Moldan. © 1997 John Wiley & Sons Ltd.

observations, first on the six countries expressly covered in the country chapters and then on broader topics, some of which are applicable to the region as a whole.

8.1 REFLECTIONS ON SIX CEE COUNTRIES

First a note on how the specific countries came to be included in the book. Inclusion of Bulgaria, the Czech Republic, Hungary, Latvia, Poland and Slovakia, was part design and part chance. The goal was to include a representative sample of countries using the most highly respected, expert authors possible. The result is a book covering the Visegrad countries, one Balkan State and one Baltic State, written by nationally and internationally regarded authorities in their fields. Though the sample size of six countries leaves out many countries of the CEE region, inclusion of countries from these three sub-regions does allow for some intra-regional comparison as well as region-wide observations. Below are selected observations on the specific countries covered, in alphabetical order:

– *Bulgaria*   In recent years Bulgaria clearly has been the centre of 'environmental' attention among the Balkan countries. The October 1995 European Environmental Ministers Conference in Sofia set the stage for very high level activities and commitments on both the domestic and international stage concerning the environment in Bulgaria and generally turned an international spotlight on the country. It is more difficult to discern, however, to what extent all the publicity and commitments are the result of enlightened leadership and sincere intentions versus the effect of national and international political inertia. Without doubt, Bulgaria is blessed with very valuable natural assets and energetic people with innovative ideas on how to protect and maintain these assets, as the county chapter illustrates. Certain serious problems exist in the country, however, such as very slow progress with economic reform, including privatization, political polarization between the Presidency and Parliament, and at least the perception of the pervasive influence of organized crime groups. The persistence of such problems will profoundly hinder Bulgaria's transition to a market-based democracy which would allow it to effectively fulfil the often impressive environmental initiatives and commitments it has expressed.
– *Czech Republic*   The Czech Republic offers a contrast with Bulgaria, and indeed many of the other CEE countries as well. In many ways it clearly has progressed further through the transition process, and in a more steady and consistent manner, than most, if not all of the other CEE countries. Some economic indicators in the Czech Republic are already comparable with those of some EU Member States, and it is not difficult to understand why many Czech persons do not think of themselves as

'Eastern' Europeans. After all, the Czech capital, Prague, lies further west than the capital of 'Western' Austria. The Czech Republic's progress in overcoming its environmental challenges reflects its advancement through the transition period. Quiet, steady, pragmatic work has led to steady and substantial improvements. Opportunities were seized as they became available, especially early after the 'Velvet Revolution', and environmental policies and laws were revised and revamped on a major scale. Nevertheless, major challenges do remain, such as restoring the extremely degraded areas of the northern coal-basins, where excessive industrial concentrations created serious environmental health hazard zones and left severely damaged ecosystems. Moreover, as the Czech Republic's economy and society become more akin to those of its Western neighbours, it will increasingly experience the kinds of environmental problems long unsolved in the West: unsustainable appetite for resource consumption; growing harm caused by growth in car usage; the generation of unmanageable amounts of solid waste, etc. The Czech people have thus far been quite successful in approximating the development styles and standards of the wealthy Western nations, one can only hope that they will be more successful in learning from, and avoiding or solving the problems they experience.

– *Hungary* In comparison with either Bulgaria or the Czech Republic, the environmental protection realm appears less dramatic in Hungary. Not being the stage for an international environmental spectacle like the European Ministerial Conference and not having environmental problems the magnitude of those existing in the Czech Republic, this should not be a surprise. Hungary probably has the fewest, and least severe, environmental 'hot spots' of any of the Visegrad countries, and thus environmental protection has not garnered the kind of attention it has in the other countries. For the most part, Hungary was spared the excessive industrialization that parts of Poland, the Czech Republic and Slovakia were subjected to and the related environmental disasters. The dominant environmental (and not only) challenge in Hungary is probably posed by the country's major urban agglomeration – Budapest. The largest metropolitan area in the CEE region, with its population of ~ 2 million it dominates the country, whose total population is ~ 10 million. The large city suffers from the environmental ailments typical of such urban concentrations – waste management, congestion, noise and air pollution from automobiles, water pollution – and will be increasingly challenged by them as these problems are some of the fastest growing throughout the region.

– *Latvia* Within the group of countries covered in the country chapters Latvia is notable for at least three reasons: (i) it is one of the smallest (only Slovakia is smaller in size) and is the least populated (with about 2.6 million residents); (ii) it seems to have less dramatic environmental problems; and (iii) it is the only country to be a member of the former Soviet Union. Latvia, like the other Baltic States, started its transformation efforts a bit later than countries of the other CEE sub-regions, and is at a correspondingly earlier stage of progress. In terms of environmental challenges the country is in the fortunate position of being without the kind of severe problems existing in many other CEE countries. Latvia's history in the Soviet Union, however, leaves it with a bit stronger legacy of Communist style central planning, and some of the difficulties associated therewith, which are described quite bluntly at times in the country chapter.

– *Poland* In terms of size and environmental conditions, one could compare Latvia to the north-east quarter of Poland, which is considered the cleanest, most environmentally healthy part of the country. Indeed the area is called the 'Green Lungs' of Poland, dominated by lakes, wetlands and forest areas, offering a respite from the environmental threats found elsewhere in country. It is difficult to compare Poland with the other countries because of its size: at $\sim 313\ 000$ km^2 and 38 million people it is simply much, much larger than its fellow CEE countries. To give some perspective, the other three Visegrad countries – Czech Republic, Hungary and Slovakia – combined together are still much smaller than Poland, in terms of both area and population. When the breadth and magnitude of Poland's environmental problems are factored in with its overall size and population, it may be fairly said that Poland faces a tremendous environmental challenge. This challenge has been recognized, both in Poland and internationally, and great efforts have been undertaken, are underway, and are being planned to address the issues. Poland has been very successful in a number of key environmental protection areas and is proud of this success: it is, for instance, widely acknowledged as having the most well-developed environmental finance system in the region. At the same time, however, the severity and urgency of some of Poland's environmental problems has yielded a rather stop-gap approach to environmental management. A more holistic, integrated yet pragmatic approach with technical, administrative and financial feasibility has been slow in coming and recent plans prepared by the national environmental authorities fail to adequately redress this weakness.

– *Slovak Republic* The contradictions and polarities of this country cannot be ignored. On the one hand it has one of the strongest, most dynamic economies in the region, while on the other a government that has been accused of becoming progressively anti-democratic and even repressive. The country enjoys wonderful areas of natural beauty, including picturesque mountain vistas and scenic valley views, while at the same time it shares the same legacy of over-industrialization as neighbouring countries Poland and Czech Republic. This picture of extremes comes through in the country chapter, which sheds a decidedly critical light on the situation in the country. The country's environmental challenges have at times become polarizing focal points, like its economic, social and political challenges, often evoking strong, emotional responses. A perfect example is the highly controversial attempt of the national government to complete construction of an old, Soviet-designed nuclear power station, with the assistance of international finance from the European Bank for Reconstruction and Development (EBRD) and private investors. After a huge national and then international campaign led by environmental organizations and the Austrian Government (the plant is located not far from the Austrian border and Vienna) showed the rationale for the plant's completion to be seriously deficient, EBRD withdrew its interest in the project, its reputation damaged. Nevertheless, despite the apparent illogic of the investment the project's proponents continue efforts to arrange financing for completion of the facility. In such a volatile setting it is truly difficult to assess objectively how Slovakia is meeting its environmental (and other) challenges, and there are real concerns that the country is actually creating more problems for itself than it is solving.

8.2 BROADER OBSERVATIONS ON THE ENVIRONMENTAL CHALLENGES FACING CEE COUNTRIES

In any analysis or discussion of the environmental challenges facing the countries of Central and Eastern Europe a number of key issues, themes, and phenomena require attention. While it is important to acknowledge that the CEE countries do have their own unique characteristics and circumstances, something often neglected in discussions about 'Central and Eastern Europe', 'economies in transition' or simply 'the region', it is likewise true that these countries share much in their recent history, are struggling with many of the same problems now and have similar dreams for the future. Some observations on issues of critical importance to many or even all the CEE countries are posited below, with the hope that they might answer some questions, provoke evaluation and new questions, and encourage new ideas and innovation, all of which can help overcome environmental problems in CEE:

– *Quality and availability of environmental information* The state of environmental information in and about CEE countries is still far from ideal and often renders meaningful analysis and confident planning impossible. The authors encountered major problems with environmental information about CEE countries during the preparation of this book and were at times frustratingly constrained by the lack of reliable, up-to-date information, analysis and data. The list of information weaknesses is long: data and information are old or simply not existing; sources are unknown or unreliable; information is inconsistent; definitions of terms and concepts vary dramatically; environmental monitoring systems are lacking, inadequate or not properly maintained; information is not easily available, etc. In a region as dynamic as Central and Eastern Europe, the timeliness of information is especially critical, with circumstances often changing considerably in the span of a year or two. The upshot of it all is that readers of this and other publications about environmental issues in CEE countries are cautioned in consuming and interpreting data and drawing very specific conclusions therefrom. (The problem of course also exists in the West, where statisticians having magical powers are often able to support opposite conclusions with the same data set, but in CEE countries the problem is more profound and pervasive.)

– *Past public exposure to environmental information* CEE countries were held in a sort of intellectual isolation for the duration of the Communist regimes. Information and concepts from the West were censored or sometimes even prohibited while the flow of information and ideas emanating from within CEE countries themselves was also suppressed and controlled. At the same time, politically honed propaganda concerning all possible subjects was released by the governments to show the people the 'real', government approved, reality. The results of this are many, but particularly important is a deep mistrust of government statements and information. The people are not sure what to believe or when, resulting in antipathy, already common as a result of 45 years of mind-numbing centralized planning. Also, with the lack of reliable information (discussed above) and widespread mistrust, decision-making is often not based on rational thinking, thus familiar approaches based on personal contacts and mutual back-scratching still persist.

– *Public participation in decision-making* This institution so critical to civil society, is starting to take root in CEE countries, though it remains a bit of a novelty and should be seen as still being in its infancy. Surely this will be one of the most important positive developments to come out of the transition period for the long term of environmental protection in the region. Recent European initiatives to establish a convention on public participation and public access to information should help to make these more than just catchy phrases and will contribute to the attainment of basic standards and practices in the region and all of Europe.

Various movements and organizations, especially independent environmental groups, could and should play an important role in promoting public participation and generally in raising environmental awareness. Indeed, there are many organized and independent environmental groups active in the CEE region (the Regional Environmental Centre estimated their number at about 1800 in 1994), and without doubt, many of these have contributed (and continue to do so) to publicizing and solving environmental problems, mostly at the local level. In numerous cases, however, environmental NGOs have been unable or unwilling to co-operate with the broader public, which can be seen as an illustration of the still immature civil societies in CEE countries. People often simply do not understand what 'constructive opposition' in a democratic society means.

– *Traditional approaches must be taken into account* Great achievements were realized in CEE countries since 1989 in establishing modern regulatory and institutional systems for environmental protection. However, in all countries of the region there remain important gaps in these systems and the transition process is far from complete. Even if completely *new* approaches and methods are introduced, it should be remembered that any *new* approaches or methods are necessarily rooted in traditional structures. This juxtaposition of new ideas and old settings is visible in practice and can create problems, especially in the areas of implementation and enforcement. Moreover, the rule of law, upon which most of these new ideas rely heavily, is not as firmly anchored in the CEE societies as in many Western countries. In order to run new regulatory and institutional systems effectively and efficiently, the people of the CEE countries must be able to understand and react to those systems. This challenging process takes years and capacity for support and implementation needs to be built.

New policies must give due respect to, and take into account, the traditional approaches and ways of thinking. While respect for traditional approaches can be critical for creating an ownership of CEE societies for solutions leading to environmental improvements, there are also old ways of thinking living strong in CEE societies which render the finding of solutions to environmental problems difficult. One good example is the discussion on energy. Previously, and for the most part still today, the debate is focused on supply and sources of energy. In addition, traditionally, there has been a tendency to think in terms of huge projects in CEE countries. But, in fact, in many CEE countries there is a surplus of power generated with inefficient and wasteful distribution systems and end use. With few exceptions the public seems to defer entirely to authorities in this issue, and there is little public involvement in the debate on energy. Fortunately, however, with the gradual rationalization of energy prices throughout the region, there is a growing appreciation of the importance and value of greater efficiency and improved demand side management.

– *Market values vs. human values* It has been emphasized at several places in the book that improvements in the state of the environment in the region and economic development are closely interrelated and interdependent. Many economic issues in the contemporary situation of CEE countries are indirectly posing great challenges for the development and implementation of environmental policies. Just a few such issues typical for the transition period will be mentioned here: the governments, the majority of citizens and many businesses in the region are experiencing severe economic hardships; a 'profit now' mentality often prevails at all levels; the public is less preoccupied with environmental issues than with other issues which have a more direct impact on daily life; with the influx of Western businesses and multinational companies the retail markets which used to be more locally based have now opened up with competition more intense than ever. Solutions for environmental protection must take into account such issues, as well as prospects for future economic development. It is a huge task to convince all stakeholders (the general public, the consumers, the producers, the administrators) that the desired rise in welfare in CEE societies *can* go (if not, *must* go) hand in hand with environmental improvements.

Another issue is the lack of general understanding of what it means to live in a market economy. People often do not fully understand, for instance, the significance of money. They do not understand that in a free-market based society money is essential for things good and bad, 'material' and 'spiritual', high and low. During Communism, money was essentially good only for things of personal consumption, otherwise it had no meaning and only the name was comparable to currencies used in market economies. It follows that many people do not believe that money is a neutral thing and need not be equated with 'materialism'. Thus a false dilemma is often perceived between 'pursuit of profit and money' against 'human values' which can impede the transition process and environmental protection therein.

– *Window of opportunity* With the collapse of the centrally planned, Communist regimes there were high hopes and expectations, especially in the East but also in the West, that the CEE countries could readily transform their political systems, their economies and their societies and rejoin their Western European siblings on a comparable footing. There was also hope that during this transition period the CEE countries would be able to learn from the successes and mistakes of the West in re-building their own new identities and societies, combining the best of East and West. It has been obvious for some time now that these hopes and expectations were largely unrealistic, not only in terms of the time required for the re-building, but also in terms of the character of the transformation.

It is true, however, that much has been gained within the region during the transition to date, and marked improvements in many areas, including environment, have been experienced. Improvements in the state of the environment should be viewed within the general progress towards more civil societies, societies which better appreciate and value the environment. (Certainly, the sharp drop in economic performance that CEE countries experienced in the first 2–4 years of transition played a major role, but this is not a complete explanation for the improvements.) These improvements are appreciated by CEE residents, perhaps sometimes more and other times less than they should be, but they are nonetheless felt. They should also be studied and learnt from in order to sustain greater improvements in the future.

This being said, it must sadly be noted that great opportunities for other, perhaps more profound, improvements have been, and continue to be, lost during the transition period. CEE countries, often under the heavy influence of external interests, are hastily adopting new products, practices, behaviour patterns, policies, laws, standards, expectations, lifestyles and cultural values which are very far indeed from the notion of 'sustainable development', so loudly touted around the globe today by Western and Eastern interests alike. Opportunities for avoiding the mistakes of the West are rarely taken advantage of by the CEE countries in any systematic or widescale manner as they rush to 'be like the West'. Unsustainable practices in all spheres (transport, resource and energy consumption, agriculture, industry, etc.) creating problems long unsolved in the West, are now appearing in CEE countries, adding to the legacies of environmental neglect and abuse of the past 50 years. Unfortunately, in their rush to 'be like the West' it appears that few of the leaders or citizens of CEE countries have bothered to really understand what 'being like the West' means and whether or not it is good for them and their countries.

The desire of CEE countries to become firmly integrated into Western organizations and alliances can also be discussed in this light. Ten countries of the region, including the six countries covered in the country chapters, have applied for membership of the European Union. The process of harmonizing of policies, regulations and institutions with the *acquis communautaire* of the European Union has already started and brings about major challenges for environmental protection. While this process seems to be leading to strengthened administrative, regulatory and institutional systems for environmental protection in CEE countries, there is the danger that the traditional shortcomings in implementing and enforcing such systems will prevail for a long time. Huge efforts from both sides, the CEE countries and the European Union, will be necessary to make the best out of this process.

– *Acute short-sightedness vs. enlightened courage* Indicative of the 'rush' mentioned above are the approaches typically employed by the CEE countries towards environmental protection during the transition period. Yes, one may hear or read phrases like 'sustainable development', 'long-term strategy' or 'integrated planning' in many public presentations and official documents, but a closer review of approved policies, and especially real actions and expenditure, will show a heavy emphasis on short-term, end-of-pipe fix-its, often carried out with little to no regard for the implications on other social or economic sectors or the overall financial or environmental efficiency of such activities. The speed of introducing new and modern systems (regulations, policies and institutions) for economic and social development is often overwhelming the capacities to implement and evaluate such systems in a comprehensive and integrated way.

It is widely agreed that a basic minimum level of environmental awareness and appreciation is lacking among CEE country residents, that an 'environmental ethic' as witnessed in some Western countries is missing. Yet, despite numerous small-scale projects and programmes in the region, expenditure on more long-term, holistic approaches to environmental protection and sustainable development must surely pale in comparison with expenditure on expensive high-technology pollution abatement equipment and facilities. Underlying this is a real need for urgent abatement efforts, but also a very basic, though perhaps not always openly expressed sentiment, that a clean environment is a nice asset to have and maintain, once other more important needs are satisfied, namely economic. (This belief persists even in the West, where it just happens that these other 'more important' needs tend to be more widely met than in the CEE countries, thus making it easier for residents of wealthy Western countries to consider themselves 'environmentalists'.) Here CEE environmental advocates have much to learn and do if they are going to be successful in convincing their fellow residents that their economic needs can only be sustainably met if their countries' natural capital and environmental futures are protected and maintained as dearly as their personal pocket-books. Let us hope that the advocates themselves recognize this and are not daunted by the formidable challenge at hand.

Index

Accident prevention 125, 159, 188
Acid deposition 78, 80, 113
Acid rain 113, 163, 195, 203, 217, 221
Aggtelek (Hungary) 145
Agriculture and environment 29, 38, 40,
    43, 45 (Box)
    Bulgaria 67–8, 74, 76, 92, 96, 103
    Czech Republic 110, 112–13, 117–18
    Hungary 133, 139, 143, 151, 157
    Latvia 169–71, 173–4, 177, 187
    Poland 198–200, 203–4, 225
    Slovak Republic 231, 246, 250
Aid, foreign see Foreign assistance
Air pollution/quality 7, 10, 30, 194
    Bulgaria 68, 70–5 (Figs), 79–81, 96–7
    Czech Republic 109, 111–16, 122, 127
    Hungary 133–7, 146, 150, 155–6,
        159–63
    Latvia 177–9
    Poland 194–7, 218, 225
    Slovak Republic 231, 233, 250
Air protection legislation 22, 39–40
    Bulgaria 86–7
    Czech Republic 116, 123
    Hungary 151
    Poland 213
    Slovak Republic 247–8
Albania 34, 41, 43, 57
Arsenic contamination 69, 76
Asenovgrad (Bulgaria) 69, 71, 74 (Fig.),
    98
Assistance foreign/western see Foreign
    assistance
Atomic power see Nuclear energy
Audits, environmental 36, 70, 88, 150,
    158, 208
Austria 4, 6–10 (Fig.), 16, 56–60, 180,
    231, 275
Awareness, public see Public awareness

Balaton Lake (Hungary) 145, 161
Baltic Council of Ministers 189
Baltic Environmental Forum 188
Baltic Sea 174, 177, 198, 208, 217, 221
Bank for Environmental Protection
    (Poland) 219
Bankruptcy 139, 144, 158–9
Batteries 153–4, 156
Belarus 189
Belchatow (Poland) 197
Beli Lom/Osam river basin (Bulgaria) 82
Beskydy mountains (Czech Republic)
    116
Best available technique/practice 81, 126,
    153
Biodiversity 10, 30, 32, 45 (Box)
    Bulgaria 77, 85, 89, 90, 101
    Czech Republic 113, 117–18, 125, 127
    Poland 203–4, 217, 220–1, 225
    Slovak Republic 251
Black Sea 29, 77, 79–83, 85
Black Triangle 194, 203, 221
BOD5 74, 82, 112, 116, 231
Bohemia 109, 111, 115–16, 119
Borrowed Nature (Bulgarian NGO) 101
Bratislava (Slovak Republic) 232, 262
Brontosaurus (Czech NGO) 128
Budapest (Hungary) 134, 273
Bulgaria 22, 28–30, 34, 37, 51–4,
    67–106, 272
    economic indicators 4–5, 13–14, 16
    foreign assistance 41–3, 57, 91, 96–9,
        103
    state of the environment 6–10 (Figs),
        70–85, 102–3
Bulgarian Society for Protection of Birds
    101
Bulgarian Union for Protection of the
    Rodope Mountains 101
Burgas (Bulgaria) 69, 71, 75 (Fig.), 83

Cadmium contamination 76, 174
Capacity, institutional 19–20, 50, 56, 58, 183, 237–9
  EAP implementation 31, 33, 36, 61, 85, 189
  see also Environmental management; Environmental institutions
Capital market, commercial 50, 55, 61–2, 169, 219
Carpathian mountains 221, 223, 229–30
Cars/car use 15–16, 158
  Bulgaria 68, 72, 87, 96–7
  Czech Republic 118
  Hungary 133, 155, 158
  Latvia 173, 178
  Poland 197
  Slovak Republic 237
  see also Transport
CEE countries 1–66, 271–80, 63 (notes)
Central planning/control 2, 274, 276, 278
  Bulgaria 68, 93
  Hungary 137, 149, 166
  Latvia 169
  Poland 193, 196, 208, 214
  Slovak Republic 230, 243
  see also Communist system; Government
Centre for Energy Efficiency (Czech Republic) 129
Cesky Krumlov (Czech Republic) 116
Charges, environmental *see Economic instruments*
Chemical substances 124, 127, 220
Chemicals industry 68–9, 81, 116, 172–3, 201, 230
Children of the Earth (Slovak NGO) 261
Club of Ecological Journalists (Slovakia) 262
CO emissions 68–9, 111, 178–9, 194, 231
CO_2 emissions 10 (Table), 111, 115, 194, 196, 217
Coal 50
  Bulgaria 72
  Czech Republic 109–10, 113–15, 273
  Hungary 137–8
  Poland 193–5, 197–8, 200–2, 220, 225
  see also Energy
Coalition Clean Baltic 190
COD 112, 231
Coke 115, 223

COMECON 48, 67, 195, 220, 246
Communist regime/system 2–11, 19–20, 23, 25, 47, 49, 275–8
  Bulgaria 67, 100
  Czech Republic 107–9, 117–19
  Hungary 131, 141, 149
  Latvia 170, 185, 274
  Poland 193, 211, 220, 222
  Slovak Republic 230, 232, 247, 262
  see also Central planning; Industry
Competition 13–14, 38–9, 169
Consultancy services 99, 163, 189, 221, 264
Consumption patterns 14–15, 120, 133, 224, 265
  see also Lifestyles
Copper 76, 174, 194, 200
Croatia 57
Czech Business Council for Sustainable Development 129
Czech Environmental Agency 120
Czech Environmental Institute 119, 120
Czech Environmental Management Centre 36, 129
Czech Geological Institute 119, 120
Czech Republic 22, 37, 55, 107–29, 221, 231, 272–3
  economic indicators 4, 5, 13–14, 16, 110
  foreign assistance 41–3, 57
  state of the environment 6–9 (Fig.), 108–9, 111–18
Czech Union of Nature Conservation 128
Czechoslovakia 28, 41–3, 107–10, 120, 248
  Velvet Revolution 110, 234, 238–9, 255, 273
Czorsztyn (Poland) 223

Danube river basin 29, 79–82, 85, 151, 229, 233
Daugavpils (Latvia) 172, 179, 190
Debt-for-environment swap 54, 62, 91, 98, 216–19
Decentralization 12, 25, 45 (Box), 48, 51
  Bulgaria 95–6, 102–3
  Poland 210–11, 215
  Slovak Republic 238, 240, 243
Denmark 28, 35, 56–60, 180, 199, 217
Desulphurization 98, 195, 197

Devnia (Bulgaria) 71
Dimitrovgrad (Bulgaria) 69, 71, 75 (Fig.)
Dobris Assessment 29
Dobris, pan-European Ministerial
    Conference 28, 253
Drinking water 45 (Box)
    Bulgaria 73
    Czech Republic 112
    Hungary 140–1, 151
    Poland 200, 210, 225
    Slovak Republic 231, 250
    see also Water quality
Dubnica river basin 82
Duha – Friends of the Earth (Czech
    NGO) 128
Dust, airborne *see Particulates*

EAP 28–34, 44, 59, 84–5, 253–5
    see also NEAP
EAP Task Force 30, 61, 63
EBRD (European Bank for
    Reconstruction and Development)
    30, 56–61, 98, 236
    nuclear power 83, 234, 260, 262, 275
    see also PPC
Eco-labelling 35, 150, 154, 226
Ecofund 62, 217–18
Ecoglasnost (Bulgaria) 79, 88, 100–1
Economic indicators for CEE 4–5, 16
Economic instruments 21, 24, 30, 32, 38,
    50, 52–3
    Bulgaria 75–7, 86–7, 93–4, 96–7, 103
    Czech Republic 116, 119, 123–4, 126
    Hungary 137, 152–8, 166
    Latvia 183, 186, 191
    Poland 194, 209–10, 215
    Slovak Republic 235–6, 241
Economic policy, integration with
    environmental policy *see Integration*
Economic reform and transition 1–2,
    11–15, 31, 41, 47–9, 277–80
    Bulgaria 68–70, 99, 102, 272
    Czech Republic 110–11, 119, 121, 272
    Hungary 132–3, 149, 153, 158–9,
        165–7
    Latvia 169–70, 191
    Poland 193–4, 223
    Slovak Republic 234–6
    see also Integration
Education, environmental 10, 36, 52
    Bulgaria 91, 95, 100–1, 104

Czech Republic 119, 122, 129
    Hungary 147, 152, 162
    Latvia 170, 187–8, 192
    Poland 208, 212, 218, 220, 222–4, 226
    Slovak Republic 238, 251, 264
EEA (European Environment Agency)
    29
Elbe river 116
Electricity 13, 83, 134, 195, 232
    see also Pricing; Energy
Eliseina (Bulgaria) 74 (Fig.), 76, 98
Energy 6, 29, 38, 43, 45 (Box), 277
    Bulgaria 67–8, 83
    Czech Republic 110, 113–15, 127
    Hungary 137–8
    Latvia 169, 172, 177 (Fig.), 187,
        190–1
    Poland 193–8, 201, 208, 220
    Slovak Republic 232, 235, 243, 250
    see also Power plant; Pricing
Enforcement of environmental law 9,
    21, 40, 45 (Box), 50, 277, 279
    Bulgaria 68, 77, 89–91, 93, 96, 103
    Czech Republic 110
    Hungary 148, 152
    Latvia 172, 182, 185, 192
    Poland 212, 214, 216
    Slovak Republic 236, 247, 264
Environment for Europe Process 28–36,
    42, 44, 59, 84, 166, 253
Environmental banks 24, 45 (Box), 62,
    219
    see also Financing
Environmental certification 186
Environmental financing *see Financing*
Environmental funds 24, 44–5, 50–5,
    61–2
    Bulgaria 76, 88–9, 91, 93–6, 98, 101
    Czech Republic 119–20, 123
    Hungary 151, 154–5, 159, 161–2
    Latvia 183
    Poland 215–9, 224
    Slovak Republic 236–7, 263
    see also Financing
Environmental Impact Assessment 21,
    30, 40, 44
    Bulgaria 85–7
    Czech Republic 123
    Hungary 151–2, 158
    Latvia 182, 186, 190–1
    Poland 206, 213
    Slovak Republic 241, 248, 259–61

Environmental information   10, 23, 32,
    67, 91, 275–6
  free access to   21, 40, 86, 120, 122,
    152
  Hungary   133, 164
  Latvia   183, 187–8, 191
  Slovak Republic   252, 258, 264
Environmental Inspectorates   24, 50
  Bulgaria   70, 88–90, 93, 104
  Czech Republic   116, 119–20, 123
  Hungary   145, 148, 159
  Latvia   182, 187
  Poland   206–7, 209–11, 213–14,
    216
  Slovak Republic   241
Environmental institutions   2, 23, 31, 42,
    45 (Box)
  Bulgaria   68, 84, 87, 89–92, 103
  Czech Republic   120
  Hungary   132, 145–8
  Latvia   170, 180–4, 187, 191
  Poland   206–11
  Slovak Republic   239–46, 262–3
  see also Government; Capacity
Environmental insurance   186
Environmental law   9, 21–2, 40–1, 45,
    49–50
  Bulgaria   68, 70, 83–4, 86–9, 91, 104
  Czech Republic   110, 123–5
  Hungary   151–3, 163, 167
  Latvia   185–7, 192
  Poland   212–13
  Slovak Republic   238, 243, 247–8,
    263–4
  see also Enforcement
Environmental management   19–20, 33,
    36, 150, 185, 189, 219
  lack of experience   12, 20, 119, 184
  *see also Capacity; Environmental
    institutions*
Environmental Management Training
    Center   98, 101
Environmental movements/NGOs *see*
    NGOs
Environmental Partnership for Central
    Europe   236
Environmental policy   19–20, 23,
    28–45
  Bulgaria   68, 70, 84, 86
  Czech Republic   120–3, 125–8
  Hungary   149–51, 164
  Latvia   184–5

  Poland   211–12, 222
  Slovak Republic   249–51, 254
  *see also Integration; Sustainable
    development*
Environmental Protection Club (Latvia)
    190
Environmental standards *see Standards*
Environmentalists   15, 17–18, 118, 128,
    131, 187, 280
  see also NGOs
Equity, green   55, 61, 63
Erosion   76, 78, 113, 151, 173, 226
Estonia   34, 37, 51, 54, 188, 189
  economic indicators   4–5, 16
  environmental indicators   6–8 (Fig.),
    22
  foreign assistance   41–3, 57
Europe Agreements   29, 37–8, 41, 84–5,
    88, 185, 192
European Commission   28–9, 38, 40,
    57–9, 89, 188
European Council   29, 38
European Investment Bank   56–61
European Parliament   43, 234
European Union accession of CEE
    countries   1, 29, 35, 36–44, 45
    (Box), 279
  Bulgaria   98, 104
  Czech Republic   125
  Latvia   192
  see also PHARE
European Union law and policies,
    harmonization of   21–2, 37–41,
    44–5, 49, 88, 192, 220
European Union   25, 28, 83, 98, 129,
    221
  fifth Environmental Action
    Programme   37, 45 (Box)
  internal market   35, 38–41
Eutrophication   81, 112, 174, 176–7
Expenditure environmental   47–8, 51–4,
    56
  Bulgaria   92–7, 103
  Czech Republic   119
  Hungary   159–61
  Poland   214–6
  see also Financing

Fertilizer   112–13, 117 (Table), 151, 199,
    246
  fertilizer industry   69, 79, 82

Fertö lake (Hungary) 145
Financial assistance, foreign *see Foreign assistance*
Financing, environmental 24–5, 29–30, 34–5, 38, 44–63
  Bulgaria 92–9, 104
  Hungary 155, 159–62, 167
  Latvia 183
  Poland 213–19, 274
  Slovak Republic 236–7, 263
  see also Expenditure; Investment environmental
Fines, environmental *see Economic instruments*
Finland 9 (Fig.), 56–60, 217, 219
Fly ash 198, 200–2, 225
Foreign assistance 10, 23–7, 36, 46, 55–63, 161, 189
  Bulgaria 91, 96–9, 103
  Hungary 158–9
  Latvia 188–9, 191–2
  Poland 216–20, 223
  Slovak Republic 236, 263
Foreign (western) enterprises 24, 161–5, 186, 219, 234
Foreign investment 5, 29, 35, 158, 169, 235
Forest (management)
  Bulgaria 78–9, 89–90
  Czech Republic 113, 114 (Table), 122, 125, 127
  Hungary 145, 149
  Latvia 169, 171, 187
  Poland 195, 203–6, 211, 213, 219–20, 225
  Slovak Republic 231, 233, 235, 250
Foundation for a Civil Society (Slovak NGO) 236
France 4, 6–8 (Figs), 10 (Fig.), 16, 54, 56–60, 91, 217
Fuel (fossil) 32, 87, 71, 114, 177
  see also Energy
Funds, environmental *see Environmental funds*

Gabčíkovo-Nagymarosdam 131, 233, 237–8, 243, 252, 261
Galabovo (Bulgaria) 71, 74 (Fig.)
Gas 72, 110, 114, 136–8, 177–8, 195–7, 219–20
  see also Energy

Gasoline 136, 151, 153–5, 161–2
  lead-content 30, 72 (Box), 118, 158
  see also Transport
Gauja National Park (Latvia) 183
GDP/GNP 5, 16, 68, 110, 133
German Democratic Republic, former 41–43, 109
Germany 80–1, 194, 199, 221
  assistance to CEE 56–60, 188–9, 217, 219, 221
Global Environmental Facility 56–61, 91, 98, 220, 236
Global environmental problems 127–8, 163, 250
Gorna Orahovitza (Bulgaria) 82
Government (national) 11, 14, 19–20, 45 (Box), 48, 70
  co-operation 17–18, 33–4, 48, 87, 103–5, 187
  responsibilities 91–2, 147–8, 187, 208–9, 211
  policy integration 17–18, 33–4, 170, 187, 209, 242–4
  see also Ministry of Environment; Local authorities; Integration
Greece 78, 80
Green Balkans Movement 101
Green Lungs of Poland 204–6, 274
Green Party 101, 190, 222–3, 245, 257
Greenhouse gas emissions 127, 217, 220
Greenpeace 128, 261
Groundwater 40
  Bulgaria 73, 77, 85
  Czech Republic 112, 115
  Poland 198, 225
  Slovak Republic 231

Hazardous waste *see Waste, hazardous*
Health human and environment 10, 28, 31–2, 40, 59
  Bulgaria 73, 76, 79, 98, 100
  Czech Republic 118–19, 122, 124–8, 273
  Hungary 147–8, 150, 152
  Poland 196–7, 210–2
  Slovak Republic 249
Heating 71, 79, 177, 195, 197–8, 219
Heavy industry *see Industry, heavy*
Heavy metal contamination 74, 76, 78, 112, 173–4

HIID (Harvard Institute for International
Development) 46, 102
Hortobagy (Hungary) 145
Hot spots, environmental 10, 273
Bulgaria 69, 71, 72 (Box), 85
Czech Republic 109, 119
Hungary 133
Latvia 171
Hungary 22, 36–7, 53–5, 131–68, 273
economic indicators 4–5, 13–14, 16,
133
foreign assistance 41–3, 57, 158–9
state of the environment 6–8 (Figs),
10 (Fig.), 132–45
Hydrocarbons emissions 69, 111, 178

IAEA (International Atomic Energy
Agency) 83
ICC (International Chamber of
Commerce) 36
IFC (International Finance Corporation)
61
Ignalina nuclear power plant (Lithuania)
187
Industry and environment 4–6, 10,
12–14, 23–4, 48–50
Bulgaria 67–77, 79, 92, 100, 103
Czech Republic 108–10, 114–16, 120
European environmental policy 32,
35–6, 38, 40, 45 (Box)
Hungary 133–9, 143–4, 149, 163
Latvia 169–70, 172–3, 178–9
Poland 193–202, 208, 210, 212, 214
Slovak Republic 230, 232, 235, 243,
250
see also Hot spots
INEM (International Network for
Environmental Management) 36
Inflation 5, 15, 49, 96, 133
Institute for Environmental Policy (Czech
Rep.) 129
Institute for Sustainable Communities
(Bulgaria) 101
Integration of environmental and
economic policies 126, 148, 153,
166, 187, 226, 279–80
EU accession 37, 39, 45 (Box)
incoherent policies 17–19, 25, 92,
104–5, 148, 153,
institutions 92, 209
see also Sustainable development

International agreements 32, 40, 84–5,
126, 188–9, 220–1
International co-operation 27–44, 83–5,
188–9, 220–2, 250, 252–5
International finance institutions 28, 45,
55–62, 189, 219, 236
Investment, environmental 25, 31–3,
35–6, 41–63
Bulgaria 68, 88, 93–8
Czech Republic 119–20, 123–4
Hungary 133, 157, 159–61
Latvia 169–70, 187, 189, 191
Poland 206, 214–19
Slovak Republic 237
see also Financing
Investment, foreign *see Foreign
investment*
Iskar river basin (Bulgaria) 82
IUCN (International Union for Nature
Conservation) 29

Jachymov (Czech Republic) 108
Jantra river basin (Bulgaria) 74
Japan 56–61, 98, 119, 161
Jaslovske Bohunice nuclear power plant
232–3, 237
Jaworzno (Poland) 197
Jelgava (Latvia) 172
Joint implementation 62
Junak (Czech NGO) 128
Jurmala (Latvia) 173

Kamchija river (Bulgaria) 74
Karvina (Czech Republic) 115
Katowice (Poland) 200–1, 233
Kiskunsag (Hungary) 145
Kolobrzeg (Poland) 200
Koniklec (Czech NGO) 128
Konin (Poland) 197
Krakow (Poland) 194, 233
Kremikovtsi-Sofia (Bulgaria) 69, 71, 75
(Fig.), 76
Kurdzhali (Bulgaria) 69, 71–3, 76, 102

Land management 38, 77, 127, 151, 157,
204
Land privatization 29, 40, 77–8, 145,
186, 235
see also Privatization

Land use planning 72 (Box), 127, 151,
208–10, 213, 244, 250–1
Landscape devastation 108, 112, 115,
117
Latvia 22, 34, 37, 169–92, 273–4
economic indicators 4–5, 13–14, 16
foreign assistance 41–3, 57, 188–9,
191–2
state of the environment 6–9, 170–80
Latvian Fund for Nature 184, 190
Law enforcement *see Enforcement*
Lead emissions 32, 69, 71–3, 76, 118,
174
Legnica-Glogow (Poland) 194
Leva Betaine river basin (Bulgaria) 82
Liabilities, past environmental 12–13,
29, 32, 35–6, 45 (Box)
Bulgaria 70
Czech Republic 125, 127
Hungary 132, 152, 158–9
Slovak Republic 249
Liepaja (Latvia) 172, 179
Lifestyles 14–15, 119, 121, 224, 279
see also Consumption
Lignite 109–10, 115
Lithuania 22, 34, 37, 52, 187–8
economic indicators 4–5, 13–14, 16
environmental indicators 6–7 (Fig.)
foreign assistance 41–3, 57
Local authorities/governments 12, 45
(Box), 48, 51
Bulgaria 76, 96, 104
Hungary 147, 159
Latvia 169, 183–4, 189
Poland 209–10, 215–16
Slovak Republic 240–1, 244–5
see also Decentralization,
Environmental institutions
Lovetch (Bulgaria) 82
Lucerne, pan-European Ministerial
Conference 28–30, 84, 166, 253

Macedonia FYR 57
Management, environmental *see*
Environmental management
Mandrensko lake (Bulgaria) 74
March river 232
Maritsa power plant (Bulgaria) 98
Maritsa river basin (Bulgaria) 74
Media 100, 164, 188, 243, 252, 257,
264

Mediterranean 80, 204
Metallurgy industry, Bulgaria 68–9, 74,
76–7, 93
Czech Republic 115
Poland 193, 201–2
Slovak Republic 230, 246
Metals, heavy *see Heavy metals*
Mining
Czech Republic 108, 110, 113, 115,
127
Hungary 137–8, 154, 162
Poland 193–4, 198, 200–2
Ministry of Environment 19, 23, 34
Bulgaria 70, 89–90
Czech Republic 119–20
Hungary 145–8
Latvia 180–4, 186–7
Poland 206–8
Slovak Republic 238, 239–44,
see also Environmental institutions;
Government
Mochovce (Slovak Republic) 234, 237,
243, 260–2
Monitoring, environmental 10, 26, 42, 45
(Box), 52, 275
Bulgaria 70, 74–5, 77, 85–6, 89–91,
93, 95–6
Czech Republic 116
Hungary 150, 156, 161
Latvia 191
Poland 38, 210–1, 218
Monocultures 78, 231
Montana (Bulgaria) 75 (Box), 82
Moravia, North 109, 115, 119
Municipal authorities *see Local*
authorities
Municipal waste *see Waste, municipal*

Nagymaros/Bös Dam *see Gabčíkovo*
National Center for Environment and
Sustainable Development
(Bulgaria) 89, 90
National Council of Environmental
Protection (Hungary) 147–8
Natural resources 6, 24, 49, 52
Bulgaria 67, 91
Czech Republic 107–8, 121, 125
Hungary 137, 149, 154
Latvia 171–2, 182, 186, 191
Slovak Republic 250–1
see also Pricing

Nature, pristine 10, 78, 109, 171
Nature protection/conservation 22,
    39–40
  Bulgaria 77–9, 89, 91
  Czech Republic 113, 122, 124, 128
  Hungary 144–5, 148, 159–60, 162
  Latvia 183, 186, 190
  Poland 203–6, 209, 213, 216, 218,
    222–3, 225
  Slovak Republic 233, 238, 248
Nature Protection League (Poland)
    222
NEAP 21, 30–4, 45 (Box), 185,
    254–5
  see also EAP
NEFCO (Nordic Environment Finance
    Corporation) 57–61
Netherlands 9 (Fig.), 174, 180
  assistance to CEE 56–60, 184, 189,
    199, 217
Newly Independent States 27–8, 59,
    169
NGOs 11, 23, 28
  Bulgaria 79, 84, 95, 99–102
  Czech Republic 120, 128–9
  Hungary 164–5
  Latvia 190
  Poland 206, 221–4
  Slovak Republic 236, 239, 252–8,
    261–2, 264
Nikopol (Bulgaria) 74 (Fig.), 79, 81
Nitrate/Nitrogen 32, 73–4, 82, 112,
    175–6, 198
Noise 22, 127, 146, 151, 156, 160–2,
    197
Nordic Investment Bank 57–60
Norway 35, 56–60
NO$_x$ emissions 7, 10
  Bulgaria 68, 79–80
  Czech Republic 111–12, 115
  Hungary 134–6
  Latvia 178
  Poland 194, 196–7, 217
  Slovak Republic 231
Nuclear energy/safety 83, 99, 101, 187,
    195, 223
  Czech Republic 114
  Hungary 138
  Ignalina (Lithuania) 187
  Kozloduy (Bulgaria) 83
  Slovak Republic 232–4, 237, 243, 260,
    275

Odra river 198, 225
OECD (Organization for Economic
    Co-operation and Development)
    27–30, 54, 124–5, 159, 212, 220
  see also EAP Task Force
Ogosta river basin (Bulgaria) 82
Oil/Petrochemical industry
  Bulgaria 69, 77, 93
  Czech Republic 112, 114, 116
  Hungary 136, 138
  Latvia 169, 173–5
  Poland 195
  see also Energy
Olaine (Latvia) 173
Opole (Poland) 197
Osam river (Bulgaria) 74, 82
Osek (Czech Republic) 109
Ostrava (Czech Republic) 115–16
Ozone depleting substances 98, 124,
    127
  Hungary 134, 136–7, 155, 157,
    165

Packaging 125, 153–6, 164–5, 179, 186,
    232
Paper and pulp industry 68–9, 116, 150,
    173
Particulates (dust) emissions 32
  Bulgaria 71–2, 75, 79, 81
  Czech Republic 111, 115
  Hungary 134–5
  Latvia 178
  Poland 194, 196, 210
  Slovak Republic 231
Paskov Biocel (Czech Republic) 116
Pernik (Bulgaria) 69, 71, 76
Pesticides 113, 117, 174, 199, 208,
    246
PHARE 37–45, 53, 56, 59
  Bulgaria 74, 79, 88–9, 91, 96, 98
  Hungary 161–2
  Latvia 188
  Poland 216–18, 220
Phosphates, Phosphor 112, 174–6,
    198
Pilis (Hungary) 145
Pirdop-Zlatitsa (Bulgaria) 69, 71, 76
Pleven (Bulgaria) 69, 71, 75 (Fig.), 82,
    93
Plovdiv (Bulgaria) 69, 71, 74 (Fig.), 76,
    93

Poland 22, 24, 27, 34, 37–8, 193–227, 274
  economic indicators 4, 5, 13–14, 16, 196
  environmental financing 51–5, 62, 213–19
  foreign assistance 41–43, 57, 216–20
  state of the environment 6–10 (Figs), 193–206
Polish Ecological Club 222
Polish German Co-operation 219, 221
Polish School of Ecologists 222
Political agendas 15, 18, 47, 170
Political reform 1–2, 11–12, 18, 26, 36–41
  Bulgaria 84, 102
  Czech Republic 110–11, 120
  Hungary 131–2
  Poland 210
  Slovak Republic 234, 263
Political support/will 25, 33, 40, 47, 50
Polluter pays principle 51, 81, 86–7, 100, 211, 263
  and revenue raising 21, 45, 86, 215
Pollution control 52, 67, 77, 193, 212
Pollution prevention 24, 52, 55, 77, 83, 86
Portugal 4, 6–8, 10, 16
Power plants 42, 72, 138
  Bulgaria 70–1, 79, 93
  Czech Republic 109
  Hungary 134–5, 137–8, 143, 159
  Poland 194, 197, 200–1, 225
  see also Energy
Poznan (Poland) 219
PPC (Project Preparation Committee) 30, 34, 59–61, 63
Prague (Czech Republic) 109, 112, 116, 119, 273
Precautionary principle 81, 86, 126
Pricing 9, 12, 67–8, 94, 141
  automotive fuel 14, 136, 237
  electricity 13
  energy and raw materials 12, 49, 115, 136–8, 193, 243, 246
  waste 12, 94
  water 12, 94, 140, 193
  see also Subsidies
Prista (Bulgaria) 81
Private (business) sector 4, 13, 35–6, 68, 126, 161, 163

environmental management 18, 20, 35–6, 45 (Box), 128
  financing 25, 51, 61
  see also State owned sector
Privatization 12–14, 29, 32, 35–6, 45 (Box), 53
  Bulgaria 68–70, 87, 95–6, 103
  Hungary 132, 138, 140, 149, 158–9
  Latvia 169–70
  Poland 208
  Slovak Republic 235, 243, 249, 263
  see also Land privatization; Liabilities
Property rights 8, 120, 124, 145, 149, 249
Protected areas *see Nature protection*
Provadijska river (Bulgaria) 74
Public against Violence (Slovak NGO) 245, 257
Public awareness/interest/pressure 10–11, 15, 29, 47, 276, 280
  Bulgaria 79, 86, 88, 95, 99–101
  Czech Republic 110, 118–19, 122, 128
  EU accession 40, 45
  Hungary 161–4
  Latvia 170, 187, 189–90, 192
  Poland 222–4, 226
  Slovak Republic 244, 250, 255–8
Public participation 10, 15, 17, 23, 26, 48, 53, 276
  Bulgaria 86, 88, 101, 104
  Hungary 148, 152, 164–5
  Latvia 190–1
  Poland 212, 226
  Slovak Republic 248, 255, 258–62

Radioactive materials 187, 232
  see also Nuclear energy
Razgrad (Bulgaria) 82
Razlog (Bulgaria) 75
Recycling 24, 112, 125, 144, 156, 165
  shortcomings 76, 124, 141, 164, 179, 200, 234
Refrigerants 137, 153–4, 155–6, 165
Regional co-operation in CEE 25–7, 29, 32, 81–5, 98, 101, 221
  see also Danube; Baltic Sea; Black Sea; Black Triangle; Transboundary issues
Regional Environmental Center 101, 236, 254, 276
Rezekne (Latvia) 179

Riga (Latvia) 172–4, 178–9, 188, 190
Gulf of 174, 176–7
Rio Earth Summit 18, 29, 205, 220, 252–3
Romania 22, 34, 37, 78–80, 82, 84–5
economic indicators 4, 5, 13–14, 16
environmental indicators 6–9 (Fig.)
foreign assistance 41–3, 57
Rossitza river basin (Bulgaria) 82
Rousenksy Lom river (Bulgaria) 74
Rousse (Bulgaria) 71, 81
Rule of law 21, 238, 241, 277
Russenski Lom river basin (Bulgaria) 82
Russia 80, 171, 178
Rybnik (Poland) 197

Salaspils (Latvia) 187
Salination 76, 78, 198, 200, 225
Sepap Steti (Czech Republic) 116
Serbia 80, 82
SEVEn (Czech NGO) 129
Sevlievo (Bulgaria) 82
Sewage/sewerage 45 (Box)
Bulgaria 73, 76, 94
Czech Republic 112
Hungary 140, 147, 151, 154, 156, 159
Poland 197–8, 208, 210, 225
Slovak Republic 231, 234
see also Wastewater
Short-term thinking and planning 14, 279–80
Siersza (Poland) 197
Silesia, Upper 194, 196, 225, 233
Silistra (Bulgaria) 81
Siltation 76, 81
Sleitere State nature reserve (Latvia) 183
Slovak Republic 22, 37, 51–4, 110, 221, 229–70, 274–5
economic indicators 4–5, 13–14, 16
environmental indicators 7–8 (Fig.), 231–4
foreign assistance 41–3, 57, 236, 263
Slovenia 22, 34, 36, 37
economic indicators 4–5, 13–14, 16
environmental indicators 6–9 (Figs)
foreign assistance 41–3, 57
SO$_2$/SO$_x$ emissions 7, 10, 30, 32
Bulgaria 68, 71–2, 74, 79–80, 83
Czech Republic 109, 111, 115
Hungary 134
Latvia 178

Poland 194–7, 210, 217
Slovak Republic 231
Social reform 1–2, 15–18, 48, 164, 169–70
economic and social hardship 47–8, 99, 119, 170, 223, 277
see also Public awareness
Socialist (Soviet) system *see Communist system*
Sofia (Bulgaria) 69, 71–2, 75, 82, 88
pan-European Ministerial Conference 28–30, 44, 84, 101, 272
Sofia Initiatives 30
Soil quality
Bulgaria 73, 76, 96–7
Czech Republic 112, 117, 122, 124
Hungary 151
Latvia 173–4
Poland 198–9, 226
Slovak Republic 250
Sokolov (Czech Republic) 115
Solidarnosc 211, 222
Soviet Army bases 112, 171–2, 180, 190, 198, 203
Soviet system/regime *see Communist regime*
Soviet Union *see USSR*
Standards, environmental 9, 32, 35–6, 39, 49
Bulgaria 68, 70, 86–7, 91
Czech Republic 110, 123–4
Hungary 147, 152, 167
Latvia 174, 182, 185
Poland 209, 212, 220
Stara Zagora (Bulgaria) 102
State of the environment 7, 23, 29
Bulgaria 70–84, 102–3
Czech Republic 108–9, 111–18, 125
Hungary 132–45
Latvia 170–80
Poland 193–206
Slovak Republic 231–4
State owned (business) sector 12, 35–6, 45 (Box), 69–70
environmental management 18, 20, 35–6, 45 (Box)
financing 25, 51, 93, 246, 263
law enforcement 110, 214, 263
see also Private sector
Steel industry 108, 110, 115, 172, 197, 232
Stonava (Czech Republic) 223

Subsidies   12, 32, 93, 117–18, 137
  to energy and raw materials   9, 49, 137
  environmental funds   53, 94, 96, 155,
    159, 161
  to industry/state owned sector   12, 25,
    68, 243, 246
  see also Pricing
Sudety mountains (Poland)   194, 203
Sumava mountains (Czech Republic)
  109
Sustainable development   27, 121, 173,
    205, 211–12, 226, 279–80
  EU accession   37, 39
  incoherent policies   18–19, 25, 92, 233,
    237
  institutions   89, 92, 208–9, 220
  see also Integration
Suwalki (Poland)   200
Sviloza (Bulgaria)   81
Svishtov (Bulgaria)   71
Sweden   56–60, 180, 184, 189, 217
Switzerland   54, 56–60, 91, 98, 104, 217
  Lucerne pan-European Ministerial
    Conference   28–9
SZOPK (Slovak Union of Nature and
    Landscape Protectors)   236, 251,
    255–8, 261

Tallinn (Estonia)   188
Tatra mountains   230, 238, 243, 261–2
Taxes, environmental *see Economic
  instruments*
Technology   12, 23–4, 52, 57, 87, 195–6,
    234
  clean technologies   14, 32, 35, 56, 99,
    218, 250
  cleaner production programmes   35–6
  see also Pollution control/prevention
Teicu State nature reserve (Latvia)   183
Tereza (Czech NGO)   128
Timber   169, 171–2, 225
TIME (Bulgarian NGO)   102
Tisza wetlands (Hungary)   145, 151
Tourism   67, 78, 173, 204, 223, 225
Trade   35–6, 38, 48, 67, 220
Training   20, 26, 32–3, 36, 41–2, 45
  Bulgaria   88–9, 96, 98, 100
  Poland   216, 218, 220, 224
  see also Capacity
Transboundary issues   32, 40, 79–85,
    101, 188, 231–4

Transition *see Economic/Political/Social
  reform*
Transportation   15, 38, 43, 67
  Bulgaria   71–2
  Czech Republic   118
  Hungary   133–4, 135–7, 147, 150–1,
    155, 157–8, 162
  Latvia   169, 171, 173, 177–8, 191
  Poland   197, 208, 225
  Slovak Republic   232, 237, 246
  see also Cars; Gasoline
Tree of Life Movement (Slovak NGO)
  251
Troyan (Bulgaria)   82, 101
Tundja river basin (Bulgaria)   74
Turkey   78, 80
Turnovo (Bulgaria)   71
Turnu Magurele (Romania)   79
Turow (Poland)   194, 197

Ukraine   80, 221
UN CSD   209, 220
UN DP   61, 220
UN ECE   29–30
Unemployment   15–16, 70, 99, 110, 170,
    258
UNEP   29, 203, 220
UNESCO   221
United Kingdom   4, 6–8, 10, 16
  assistance to CEE   56–60, 91, 98
Urban areas/issues   15–16, 67, 133, 197,
    219–20
  urbanization   7, 67, 171, 172–3, 232
US AID   72, 98, 102
US EPA   98
USA   35, 119, 219
  assistance to CEE   54, 56–60, 91,
    98–9, 217
USSR, former   6, 108, 170, 173–4, 194,
    274
Usti nad Labem (Czech Republic)   221

Vah river (Slovak Republic)   232
Valmiera (Latvia)   179
Varna (Bulgaria)   71, 83
Vavrousek, J.   249
Veliko (Bulgaria)   71, 75 (Fig.)
Ventspils (Latvia)   173, 179
Vidin region (Bulgaria)   76
Vidzeme (Latvia)   183

Vilnius (Lithuania)   188
Visegrad countries   42–3, 55, 63 (notes),
   273
Vistula river   198, 225, 233
Vit river (basin) (Bulgaria)   82, 93
VOC emissions   134, 136–7
Vratsa (Bulgaria)   74 (Fig.), 75 (Fig.), 82

Warsaw (Poland)   200
Waste, hazardous   38, 40
   Bulgaria   73, 76–7
   Czech Republic   112, 124, 127
   Hungary   143–4, 147, 156–7, 158–9
   Latvia   172, 180, 186
   Poland   200–2, 225
   Slovak Republic   232, 234
Waste management   8, 12, 14–15, 24
   Bulgaria   73, 76–7, 94, 96–7
   Czech Republic   109, 112, 122, 124,
      127
   Hungary   140–4, 146–7, 157, 159–63
   Latvia   179–80, 189
   Poland   200–2, 208, 210, 215, 218, 225
   Slovak Republic   230, 232, 238, 241,
      250
Waste management legislation   22, 39, 40
   Bulgaria   76
   Czech Republic   112
   Hungary   151
   Poland   213
   Slovak Republic   247
Wastewater (treatment)   8, 12, 32, 42, 45
      (Box)
   Bulgaria   74–6, 82, 87, 93, 96, 98
   Czech Republic   109, 112, 116
   Hungary   140–1, 151, 157
   Latvia   171, 174–5, 189
   Poland   197–9, 202
   Slovak Republic   237
   see also Sewerage; Water pollution

Water legislation   22, 39, 40
   Czech Republic   124
   Hungary   151
   Poland   211, 213
Water pollution/quality   9, 38, 40, 45
      (Box)
   Bulgaria   73–7, 79–83, 88, 96–7, 101
   Czech Republic   109, 112, 122, 127
   Hungary   139–40, 141, 146–8, 151,
      156, 159–63
   Latvia   174–7, 189
   Poland   197–200, 208, 211, 214–16,
      218, 225
   Slovak Republic   231, 241, 250
   *see also Drinking-/Ground-/Waste-
      water; Sewerage*
WBCSD (World Business Council for
   Sustainable Development)   36
Western enterprises *see Foreign enterprises*
Western assistance/aid *see Foreign
   assistance*
Western lifestyles *see Lifestyle*
Wilderness Fund (Bulgarian NGO)   101
Win-win solutions   17, 32, 166
Window of opportunity   30, 278
World Bank   56–61,72, 76, 83, 98, 219
   EAP/NEAP   28, 31, 34
WWF International   190

Yantra river basin (Bulgaria)   82
Yugoslavia, former   41–3, 76, 78–9

Zahorie lowland (Slovak Republic)   230
Zeleny kruh (Green Circle) (Czech
   NGO)   128
Ziar nad Hronom (Slovakia)   237, 243
Zielona Gora (Poland)   200
Zinc   76, 174, 200
Zlatitsa (Bulgaria)   71, 74 (Fig.)